The Boys Book of Sea Fights Volume 1

Great Battles for Boys Teen Series

Chelsea Fraser

The Boys Book of Sea Fights, Volume 1

Print Edition

TABLE OF CONTENTS

FOREWORD

THAT THE SEA has furnished its full quota of thrilling and important conflicts in the history of mankind, standing a close second in this wise to the land, there can be no doubt.

The consensus of opinion among students discloses it. The archives of nations prove it.

Moreover, for spectacular exploits of both individuals and small bodies of men, the sea has always been a leader. Its very structure is a constant threat of death or stirring adventure, presented at the most unexpected moments;.

And when to this peril is added the chance of encountering hostile ships, we cannot wonder that sailors have performed deeds of surpassing heroism and accepted supreme sacrifices for the sake of their country that thrill us to the very core of our bodies.

This book tells the tale of sailors at war.

Its heroes are real heroes—men who have lived, who have performed the very deeds recorded of them, so far as careful research can verify.

They are not the sailors of any one country or nation, but of many countries and nations—which fact, it seems to me, really makes them no less brave and their deeds no less worthy of recounting.

Valor and patriotism, no matter by whom displayed, surely ought to meet with our admiration and stir us to like deeds for our own people and our own country.

C.C.F.

SECOND FOREWORD

CHELSEA FRASER'S *The Boys Book of Sea Fights* is a thrilling read.

But it's also a bit challenging, which is precisely why I hope you will read both Volume I and Volume II of this book.

You will learn some amazing naval warfare history. But you will also become a much stronger reader and smarter person for staying with the books, even if you have to read slowly to absorb all the exciting details.

To help you with the vocabulary and nautical terms that might seem unfamiliar, I've added brackets—they look like this: [ ]. Inside the brackets are definitions and other explanatory notes.

However, I've left some of these terms alone! YOU look them up! Not only is it a good practice, you will develop the habit of keeping a dictionary nearby as you read. The best readers do that.

Also, there are many foreign places mentioned in these two volumes. Early sea explorers sailed across the planet, and wars were fought in distant colonies. For that reason, you will want to have a map or globe nearby as you read. Find the places he mentions—you will become a geography whiz!

And I promise, you won't regret these habits. The knowledge you gain will only make you appreciate even more the heroic feats of these brave warriors.

(You'll also know a lot more about sea battles than most adults!)

Now, let's dive into Volume I of these fascinating biographies and naval battles described by a very popular writer of his time, Chelsea Fraser.

Joe Giorello

CHAPTER I

SIR FRANCIS DRAKE

Portrait of Sir Francis Drake by W. Hall, 1604

Great golden galleons plied the Main;
He hit them hard, again, again;
And with his fleet of British ship
Forever broke Spain's tyrant grip.

—*Fiske.*

Early Adventures

THE STIRRING TIMES into which young Drake was born acted as a forcing-house for the growth of character.

Boys turned into men at a leap. Bred in a nursery whose very atmosphere was war and revolution, they were trained by danger and privation to fight battles at an age when the boy of today is making ready to take up his school life.

The strange mixture of lax moral standards and fierce religious passions, the prevalence of an unwonted bigotry, the light esteem in which human life was held, the rapid succession of startling events, the persecutions carried on in the name of a holy cause—all went to forge men of singular and violent contrasts.

Sir Francis Drake, the foremost sailor of the Reformation [side note: more on the Reformation later], the chief pirate of Queen Elizabeth, was one of these men.

No name in England's annals of the sea has been surrounded with so dazzling a setting of romance as his.

During his whole lifetime, Drake's adventures found no place in sober history. They invaded the realm of folklore and took a strong hold on the popular fancy in the shape of marvelous tales and legends.

Yet there was a solid foundation to them for all of that, and rising out of this wonderland of romance the record of mankind must ever show that this daring boy of Devon [a county in the south of England] was one of the most skillful of navigators as well as the first really great admiral in the development of modern naval science, which had its cradle in England, and which substituted the sailing-navy for the ancient, rowing-navy.

Born in 1544, in the heat of the strife between Catholics and Protestants, little Francis held a fierce hatred for Spain and her subjects. His father, Edmund Drake, was one of the most zealous of Protestants, with a gift for preaching which he often used. He made no effort to hide his religious attitude, although Devon was known to be mainly Catholic, but went about boldly and fearlessly asserting his views.

This action could have but one outcome. The Catholic faction made it so hot for all Protestants in Devon, and for Edward Drake in particular, that he fled with little Francis to St. Nicholas Island, in the harbor of Plymouth. He then moved to Chatham Reach, on the other side of the country, beyond the dockyard, at the mouth of the Medway, which was the anchorage of vessels out of commission, of warships, and of old and useless hulks.

Here, the Protestant preacher was given an appointment under King Edward as "Reader of Prayers to the Royal Navy" and was assigned a rotting hulk as a dwelling place.

But to young Francis, it was no rotting hulk—rather, a palace of the most intoxicating delights. Seldom did he care to go ashore.

All day long, he climbed the masts of his new floating home, raced across the old weather-beaten decks, or delved into the below decks' mysteries of the veteran craft.

And when night would come he often fell asleep in the midst of his beloved old warship's guns, rocked by the heaving waves and solaced into the last deep oblivion by the lisping tide and the lullaby of the sailors' songs.

Edmund Drake had hoped to place his boy in the navy, with the patronage of Edward VI and the powerful Earl of

Bedford, but his expectation was shattered by a rude change. King Edward died.

"Bloody Mary," the Catholic Queen, succeeded to the throne, and the land was threatened with a prince of Spain as husband to the Queen.

Portrait of Queen Mary Tudor of England (1516–1558), also known as "Bloody Mary," by Antonis Mor (1519–1575)

Soon it was that the gathering storm of the Reformation suddenly burst and threw all of England into turmoil.

[The Reformation is a really big event in history. During the 16th century, Christians began rebelling against the Catholic Church's complete control of Europe's religious and cultural rules. These Christians wanted to break away from the Catholic Church and the Pope and form a more direct religious relationship with God and Jesus Christ. Eventually the Reformation divided Europe and England in deadly ways. For instance, the Catholic Queen of England, nicknamed "Bloody Mary," killed many of these protestors—that's how their religion became known as *Protest*ant. The Reformation is among the most important developments of Western civilization. I hope you will read more about it. For now, let's get back to Drake.]

Francis's father lost his position and was forced to apprentice his son as a ship's-boy on a craft that carried on a coasting trade with France and Holland. It was on this channel coaster that the sailor lad, exposed to great privations and severe conditions, gained some of the most valuable lessons of his experience.

While his young body was being steeled to every conceivable form of hardship, his spirit was being trained for future revenges on the Spanish Main. Passionate tales of the horrors of the Inquisition were being told by Flemish refugees on quay and shipboard.

[The Inquisition is *another* fascinating bit of history related to the Reformation. The Catholic Church wanted everyone to follow its religious rules very strictly. They would also investigate people who held different beliefs or practices, especially people who practiced different forms of Christianity—those Protestants again. Imagine if you and your family went to a certain church and believed in a specific way, but your neighbor believed something a bit different. During the Inquisition, your family might get questioned by religious authorities, sometimes very harshly, to find out if they believed *exactly* what the church leaders said they should. If you didn't, you would be punished, sometimes very severely.]

The Spanish Inquisition, in which heretics—people refusing Catholicism—were sometimes burned at the stake. (Artist and date unknown)

The persecutions ordered by Philip, King of Spain, spread throughout the Netherlands and fanned the flame for the English Reformation. [More and more people felt outraged by the Catholic persecution of Protestants.]

Francis Drake found himself in the center of the hottest frenzy of religious feeling. Small wonder that he grew up with such an implacable [stubbornly angry] hatred of the very name of Spain. That hatred furnished the motivating power of his subsequent brilliant career.

In the meantime, events followed one another rapidly.

Bloody Mary passed away, and the Protestants were once more in favor. Francis had grown from boy to youth; his master skipper had died and bequeathed him his little craft on which to begin life as an independent trader.

DRAKE.

Open war with Spain had not yet been declared, but the rupture was imminent [about to happen very soon]. Cruel reprisals by private individuals on both sides were rapidly paving the way for the coming break.

The English Channel swarmed with sea-rovers. Four hundred adventurers swept the narrow stretches of water in search of plunder. The enormous wealth of the Spanish trade courted depredation in those loose-moraled times, when it was not the fashion to be as strictly honest with the belongings of one's neighbor as it is today.

Spanish galleons were chased and scuttled. Catholic vessels of other nations, particularly France, were considered legitimate prey for looting. Rich cargoes of saffron, cochineal, wool, silk, gold, silver, pearls and precious stones, linen, tapestry, and wines were carried off to the pirates' lairs on the Isle of Wight and in the creeks and inlets of the Irish coast.

In revenge, British ships were seized in Spanish ports, and many English sailors were thrown into the dungeons of the infamous Inquisition.

Forced to quit his independent trading on account of its unprofitableness, young Drake entered the only service that then seemed open to him—that of his famous kinsmen, Captains John and William Hawkins, the rich ship owners and pirate merchants of Plymouth.

British merchant ship at anchor

The chief seaport town of picturesque Devon might well have been called one of the pirate centers of the English coast.

Its harbor was large and safe, and many precious cargoes, obtained by foul means as well as fair, were brought in by daring adventurers who had scoured the limited sea in search of riches.

Neither gold nor excitement were hard to find in those days, and Plymouth became, by reason of its peculiar accessibility and additional virtues, a favorite port for the bold crews.

In Queen Elizabeth's time neither the navy nor the maritime commerce of England were established on a regular footing. The navy, used simply as an adjunct to the army, had remained an undeveloped independent instrument of national power, and the vastness of its resources were still undreamed of. Later on, Drake proved the wonderful possibilities of a strongly equipped fleet in wartime by turning the enemy's coastline into the center of hostilities and by destroying his trade with foreign countries.

In October 1567, Drake set sail from Plymouth Harbor as a pilot to John Hawkins. In the squadron were six vessels, all armed and victualled [supplied with food] for a long voyage.

But meanwhile, the fleets of war and commerce were intermingled.

In times of peace, Queen Elizabeth used her men-of-war for merchantmen, while in wartime, she added her real merchantmen to her force of fighting craft.

In this way it became exceedingly difficult to draw the line between official naval expeditions and private commercial enterprises, the last-named of which, as already shown, were frequently synonymous with buccaneering [pirating].

This state of affairs worked to the disadvantage of her adversaries and much to the advantage of the Queen at times. On other occasions, the benefits were reversed. Altogether it proved an ill arrangement.

Often, the Queen was one of the shareholders in the filibustering expeditions of Hawkins and Drake and contributed some ships-of-war to the outfit. She was too much allured by the prospect of untold riches in gold, silver, pearls, and precious stones to resist the temptation of enriching her private coffers by becoming a secret partner in the buccaneering ventures of her favorite pirates.

However, when Spanish ambassadors and the diplomats voiced their upset with pirates, she was careful to assume indignation and disapproval of the practices of her unruly subjects. And while she always welcomed the precious cargoes brought back to her by her partners in adventure, the men themselves were not always sure of her approval.

Such was her vacillating trend of mind that one time she might bestow upon the returned hero the honors of knighthood, and the next time send the poor fellow to the gallows.

But the game was worth the effort, and so it was that on that October day in 1567 Hawkins and Drake sailed out of Plymouth Harbor with what amounted to a naval squadron, loaded with ammunition and numerous heavy guns.

Accompanying them were two ships-of-war, the *Minion* and the *Jesus*, especially loaned by Elizabeth herself.

Drake appeals to Queen Elizabeth: "So please your Majesty, to singe the King of Spain's beard; it has grown somewhat too long." Artist unknown.

From the very first, the squadron met with bad weather.

Near Cape Finisterre a violent storm damaged and scattered many of the vessels, but they succeeded in reaching their first rendezvous at the Canary Islands. From there they sailed southward to the coast of Guinea, in western Africa, where the traders spent several months in collecting slaves. [Side note: African tribal leaders sold their fellow black Africans to white traders, fueling one of the most shameful events in history: the slave trade.]

Partly by means of the sword, and partly by exchange of scarlet coats and beads, they succeeded in storing away in their holds as many as five hundred of these black men. Thus equipped with trading material, they crossed over to the Caribbean Sea, reaching the West Indies in the following March.

As traffic had been forbidden in the colonies by the Spanish government, they found it no easy matter to dispose of their cargo. Rio de la Hacha, on the Caribbean coast of Colombia, brought them up with a start by presumptuously firing upon the merchant fleet.

To this rude greeting Hawkins and Drake retaliated with equal severity. They blockaded the port, stormed the defenses, and carried the town by impetuous assault. Then, in secret and under cover of the night, the unlawful trade began, and two hundred slaves were exchanged for gold, silver and pearls, sugar, and hides.

Well pleased with their valuable cargo, the fleet's leaders decided to sail at once for home.

Unfortunately, however, they had tarried a bit too long in those treacherous waters. Two fierce *furicanos,* or hurricanes, disabled the squadron and compelled them to seek shelter in San Juan de Ulua, or Vera Cruz, the port of the City of Mexico.

In the harbor, the adventurers found a Spanish merchant fleet of twelve vessels. These ships were unarmed, and laden with the year's produce of the West Indies, amounting to more than ten million dollars in gold and silver. The Spaniards lay in the harbor, hourly expecting an armed escort of their countrymen to convoy them on the home-bound voyage.

Indeed, the very next morning, following Hawkins and Drake's arrival, the armed escort appeared upon the scene, coming grandly forward just outside the harbor.

It was an extremely difficult position in which the English traders now found themselves.

Hawkins was obliged to choose between two doubtful courses. Either he must appeal to Spanish honor (a very unstable virtue) in order to let him proceed on his way without attack, or he must do his best to keep the enemy's ships from entering the harbor by sheer force of his own feebler armament until the first tornado might cripple the Spaniards sufficiently to permit of him boarding them, seizing their treasure, and making off for "Merrie England."

In his quandary, Hawkins sought the opinion of Francis Drake.

The bold Drake at once advised an attack; but Hawkins demurred.

This meant the loss to Spain of more than twenty-two million dollars; it would undoubtedly precipitate the long-threatened rupture between that country and England, and while the Queen could be counted on as gladly receiving almost any ordinary treasure, Hawkins was much afraid of her displeasure at so stupendous a venture as Drake advocated.

Finally, Hawkins decided they should choose peaceful path, if possible.

He appealed to the Spanish fleet commander for a condition of friendliness. This was graciously promised.

But it was all deceit; in spite of their sacred oaths and solemn pledges, no sooner were the wily Spaniards permitted to enter the harbor before they fell upon the British ships with their superior numbers.

In the sharp but brief engagement that followed the British seamen fought with desperateness and wonderful valor. But being wholly unprepared for the dastardly attack, they could save but few of their vessels.

The smaller craft were quickly sunk, and the *Jesus* was so shattered that it had to be abandoned with all its precious spoils.

The *Minion*, with Hawkins on board, and the little *Judith*, with Drake, alone escaped on that fatal night.

Riddled with Spanish shot and terribly damaged, with crews depleted and half-starved, these two ships struggled homeward and eventually crept into Plymouth Harbor without a remnant of the former immense and valuable British cargo.

Plundering the Spanish Main

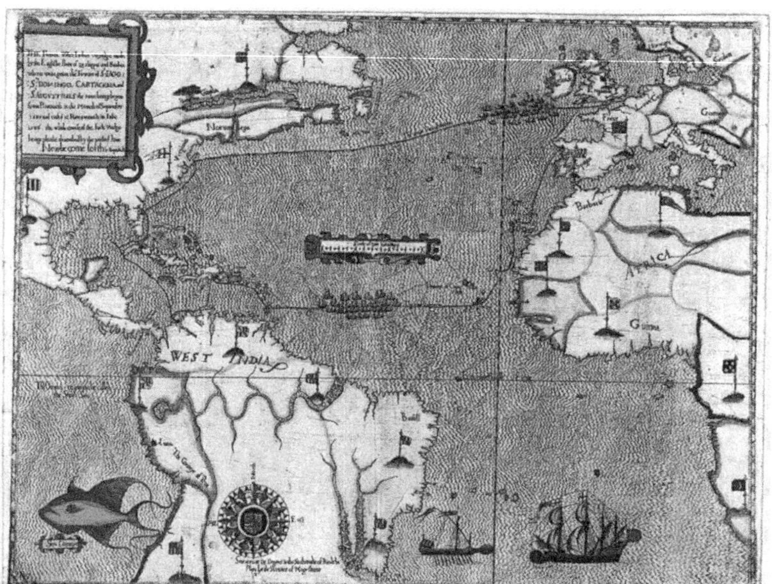

Map from 1585, showing Drake's voyage. Europe is to the upper right, western Africa below that, northern South America at the bottom, and eastern North America on the left, including Florida.

JUST THREE YEARS LATER, in 1570, Francis Drake succeeded in obtaining a sailing commission from Queen Elizabeth as a reward for his sea exploits.

During the next two years, he cruised as his own commander in the West Indies, enriching himself with much plunder.

Drake was a very successful pirate!

In 1572, on Pentecost Sunday, the Christian holiday that marks the descent of the Holy Spirit upon the apostles of Jesus Christ, Drake set sail for Plymouth at the head of a tiny squadron and a handful of men.

Leading was the *Pascha*, Drake's flagship, a vessel of seventy tons. The rear was brought up by the *Swan*, of twenty-five tons' burden, captained by his brother, John Drake, who had taken up a sea-faring life. These small men-of-war ships were fitted out with every warlike device of that time.

In addition to a plentiful supply of guns and ammunition, boarding implements, and cutlasses, there were three small pinnaces. [Light sailing ships used for trade and military operations, pinnaces were typically narrow and sleek with just one or two masts and sails. These versatile vessels could be used for reconnaissance, transport of goods or passengers between larger ships and shore, raiding, or pursuit of an enemy due to their quick maneuverability. Pinnaces were often carried aboard larger ships to be used later during boarding actions, patrols, and other tasks requiring smaller vessels. Pinnaces played significant roles in naval warfare during the Age of Sail—kind of like the PT boats of their time!]

An example of a pinnace at sea

Drake's two crews comprised seventy-three men, of whom only one was over thirty years of age. It seemed, truly, like a boy's crazy venture.

Favored by the winds, the squadron sailed without a stop until it had its first sight of Guadeloupe, one of the leeward Islands in the West Indian group. On reaching Port Pleasant, a small landlocked harbor in the Gulf of Darien, on the mainland, Drake dropped anchor and started to set up his pinnaces. It was a safe bay, and convenient for his purpose.

While he was in the midst of this work, a strange squadron hove in sight. To his great relief, this proved to consist of a vessel belonging to Ned Horsey, the well-known pirate of the Isle of Wight, and a Spanish caravel and a shallop that its commander, Captain Ranse, had captured. The two adventurers greeted one another joyfully and decided to join forces forthwith.

Seven days later, these united squadrons crept out of the harbor.

Westward along the coast they stole, keeping a sharp lookout for Nombre de Dios, the treasure house of the Spanish Main.

A week later, the comparatively small force lay at midnight under the huge bluffs at the point of the harbor of this town. Breathlessly, the British lads awaited the breaking of dawn—the time appointed for the audacious attack. Twenty-four of them were armed with muskets. The remainder were possessed of pikes and bows and arrows, with the exception of four who had been provided with drums and trumpets to inspire the crews and terrorize the natives.

Twelve men were left to hold the pinnaces, ensuring a safe retreat in case of necessity. The rest of the company was divided into two groups and advanced upon the Plaza from different sides.

Half a dozen fire-pikes lighted the way, casting a lurid glow over the narrow streets; the drums and trumpets sounded with maddening din, appearing to be the portion of a large marine band rather than of a few energetic men.

Quickly the Spanish town aroused itself. The great bells in the Catholic church and monastery began to clang out their brazen alarm. People, half-robed, appeared upon the streets and ran hither and thither, with cries and shouts that merged from fright into a threatening roar. The Spanish soldiers had been called to arms in an incredibly short time; at the end of the Plaza, near the Panama gate, they were soon drawn up to receive the attack.

As they came undaunted forward, a heavy volley of bullets and arrows greeted Drake and his force full in the face. Pausing only long enough to return the fire in like manner, they rushed on, shoulder to shoulder, and closed in upon the consternated Spanish soldiers with pike and cutlass.

Flashing the sputtering, weirdly flickering fire-pikes in their very faces, yelling like a horde of hungry beasts, thrusting to right and left with vicious jabs of sharp-pointed steel, it is not to be wondered at that the superstitious Spaniards went into a panic and presently broke and fled before this enemy they scarcely could see in the darkness and whose weapons seared and lacerated them so terribly.

Engraving of Sir Francis Drake by Jodocus Hondius, 1577. Each object symbolizes Drake's achievements. Can you find each clue?

In a very brief time, the last of the enemy soldiery and citizenry had disappeared through the Panama gate, leaving the Plaza in the hands of the adventurers.

Drake at once placed a guard at the entrances to the town.

With the rest of his men, he took possession of the governor's house. There, in a lower room, a blaze of treasure met their eyes—a blaze of such proportions that it made the bold commander himself, used to the sight of riches, stare for some moments unbelievingly.

Piled against the wall and glinting in the beams of light that came through the partly open door were great bars of silver. These reached fully seventy feet from end to end of the long room, and rose, like a huge woodpile, till the upper bars all but scraped the ceiling. The poor Devon lads, with their commander, looked on this unaccustomed sight in half-dazed wonder.

At this moment some of the Englishmen left on guard came running up with the report that the pinnaces were in danger of being captured.

Drake immediately dispatched John Oxenheim to reconnoiter the shore, and made a rendezvous at the treasure-house of the King which had not yet been visited and which stood near the water's edge. Here the chief pirate declared they were likely to discover far greater wonders in the shape of gold and precious stones—enough, probably, to overflow their pinnaces should they attempt to take it all.

But scarcely had they departed when one of those furious tropical storms of the region burst suddenly over their heads. The thunder rolled and crashed; the rain fell in torrents. Before they could get under shelter, their bowstrings had been so wetted that they were temporarily useless, and their powder ruined.

This so changed the aspect of matters that the men, losing their nerve, declined to risk a counterattack by the Spanish in the proposed visit to the King's treasure-house. Drake pleaded, then taunted.

"I have brought you to the treasure-house of the richest country in the world," he cried; "blame no one but yourselves if you go away empty!"

They stood before the treasure-house as Drake spoke.

But still his followers hesitated.

Quite losing control of himself, Drake ordered them to break into the structure or suffer the displeasure of the Queen upon their return. He took a step forward himself at this juncture. But, as he did so, he stumbled and fell on his face, blood gushing from a wound in his leg. He had been shot early in the encounter, and had concealed the fact from his men that they might not lose heart.

Quickly they lifted him from the ground, and against all his entreaties carried him to his boat. In order to preserve their beloved captain's life, they set sail, abandoning the rich spoils they had come so far to seek.

But the gold and silver bars left back there in Nombre de Dios were not forgotten, nor given up for good.

Back in a hidden bay in the Gulf of Darien, his favorite secret retreat, Francis Drake soon recovered from his wound, and made plans for stranger and more daring projects than he had yet brewed.

With the aid of the Maroons, a savage tribe of escaped slaves, he planned to intercept the gold of Panama as it was carried in mule packs across the Isthmus to be shipped to Spain. But months must elapse before the coming of the dry season, when the Spaniards were accustomed to make this annual journey overland, and meanwhile his pinnaces stole from the harbor and plundered passing ships and raided the neighboring coast.

Under these circumstances, it was not long before the name of *El Draque* [Spanish for "The Drake"] became a dreaded one to the ear of the average Spanish seaman.

Castilian captains [Castile is a region in Spain] feared to operate their ships over the Main when it was rumored the famous English navigator was in their waters. Castilian crews trembled, half the fight out of them, when this terrible crusader of the seas suddenly hove in view with his ships. Coast towns, especially Nombre de Dios, put additional guards over their treasure-houses.

To all of this strenuous effort at protection, Francis Drake only smiled grimly and continued his successful operations.

On the morning of the 1st of April, a mule train laden with gold and silver was traveling cautiously along the road to Santa Cruz.

A mile from the town, within earshot of the carpenters working at the docks, the thicket brush held a menace. But the Spaniards came on blissfully unconscious of that fact, the mule bells tinkling regularly and peacefully as they had been right along.

Suddenly the plodding sound of the many hoofs of the animals was broken by a frightful din.

Figures had unexpectedly risen beside the path in two different places, seizing the foremost and hindmost mules, which immediately had set up a terrific braying, spreading terror to those between and occupying the attention of most of the guards to keep them in restraint.

In the midst of all this, a third body of strange men rose up and discharged a heavy fusillade of bullets and arrows into the ranks of the Spaniards. There was a feeble reply in defense, whereupon the escort fled in a panic, leaving the mules and their precious burdens in the hands of the attackers.

Needless to say, these were the lads from Devon.

Swiftly the bars of silver, too heavy to carry away then, were hidden in the burrows or buried in pits especially scooped out for them. The gold itself was stowed away in shirts and pockets, and with forced marches, the young Englishmen returned to Rio Francisco.

But to their dismay, their pinnaces were nowhere to be seen!

Instead, seven Spanish ships rode menacingly in the harbor.

Spanish galleons of the 1600s that carried precious gems and gold

All hope of safety was gone.

However, Drake's ingenuity found a way out of the awkward dilemma.

A raft was built from drifted tree trunks and limbs. With a biscuit-sack for a sail, and a long sapling for a rudder, Drake and three of his men started on a wild sail over an angry sea. A strong wind lashed the waves high, many of which, as they sat for six hours in water up to their waists, broke completely over their heads.

By mid-afternoon the wind died down, and with it the tumultuous waters. Then the hot sun came out and beat hotly down upon their unprotected bodies until their wet clothes fairly steamed and their strength was well-nigh exhausted.

Fortunately, as night settled down, they managed to work their way into a quiet little cove, where, as hoped, the pinnaces were found to have sought shelter with their guard crews.

That same night, tired as he was, Drake showed his indomitable willpower and endurance by rowing back to Rio Francisco. The remainder of the crew were told of the happy finding of the pinnaces, whereupon it was decided to return at once to Santa Cruz and retrieve the hidden silver bars. This was accomplished without further misadventure, and the little company then rejoined those aboard the pinnaces.

A fortnight later the lads from Devon started on their homeward journey. They were laden with an unusually rich bounty, for, besides the treasure of the mule train, they had been lucky enough to overhaul almost two hundred vessels in the Caribbean Sea.

The voyage back to England was accomplished in due course, after a number of additional captures of Spanish ships which added in no mean way to their already enormous amount of treasure.

It was on the Sabbath Day, August 9th, 1573, that they cast anchor at last in Plymouth Harbor.

News of Drake's arrival and wonderful exploits spread like wildfire. So anxious were the townspeople, who were at divine worship in church, to see his ships and pay homage to their successful countryman, that many of them forsook their devotions to hurry down to the quay.

The Riches of the New Ocean

ON A CERTAIN DAY in November 1577, we find Drake standing on the deck of his ship, his face turned toward the fabled Pacific. Men of the time tell us that he was dressed in a "loose, dark seaman's shirt, belted at the waist, with a scarlet cap adorning his flowing black locks."

If this be true, as undoubtedly it is, he must have looked the typical bold rover of the seas that he was indeed.

Drake's ships were little cockle shells compared to vessels of modern times. The ships consisted of the *Pelican*, one hundred tons; the *Elizabeth*, eighty tons; the *Swan*, fifty tons; and the *Christopher*, a pinnace of fifteen tons.

The first two ships carried sixteen guns, while the *Swan* was a provision boat. One hundred and fifteen men and fourteen boys manned the craft, which were well ammunitioned with bullets, wildfire, chainshot, muskets, pistols, pikes, cutlasses, and bows and arrows.

[Chainshot consisted of two small cannonballs or iron balls connected by a length of chain or rope. When fired from a cannon, the chainshot would spin rapidly, creating a deadly horizontal arc as it traveled through the air. The purpose of chainshot was to damage the rigging and masts of enemy ships. When fired at the sails or rigging of an opposing vessel, the chain's swinging motion would entangle in the ropes and rigging, causing significant damage and impairing the enemy's ability to maneuver effectively. Chainshot was particularly effective in naval battles, as disabling an enemy ship's rigging could render it immobile and vulnerable to capture or destruction.]

Drake was about to sail with this little squadron into a chartless and unknown ocean, brave a shadowy world of water that popular superstition had peopled with every conceivable terror of elements and every dreadful form of animal, serpent, and fish life.

To his crew, his final destination was kept a secret. Led to believe that they were bound for Alexandria, Egypt, it wasn't until they reached the coast of Morocco that the real object of this venture was made known.

Beset almost at the start with bad weather—through stress of gales, fogs, calms, and tempestuous seas. They were also embroiled in a possible mutiny, treason, and the tragedy of an ocean lynch-court [a crew member had committed a crime punished by hanging], which was followed by an execution off the lonely coast of Patagonia, the sorely assailed fleet nevertheless kept courageously on its perilous course.

A spirit less unflinching than Drake's would have quailed under the torment of Nature and the inconsistency of man.

Only three of the five ships were brought finally to the gateway of the great South Sea. Meantime, the *Pelican* had been rechristened the *Golden Hind*, and on the 6th of September, she was the first of her sister craft to pass through the Straits of Magellan, amid cold and sickness, and enter the famed South Sea. [Find these places on your globe or map!]

Really, the dangers of the voyage had now just begun.

No sooner had the expedition entered the confines of the mighty Pacific than all the furies of a violent tempest burst over them.

For six weeks they were tossed to and fro like pebbles. Battered by the gigantic waves, the rigging and sail torn by the terrific winds, they were swept fully six hundred miles out of their course.

During the third week, the *Marigold* went down with all on board.

A week later the *Elizabeth* became separated from the *Golden Hind*, and, losing heart, its commander, Captain Wynter, returned through the Straits the way they had come, and sailed back with his sadly crippled vessel to England, there to report that in all probability his ship was the only survivor of the squadron, and to recount the awesome fury of the strange new waters which were "invested with all the wrath of Satan."

Meantime, Drake was left alone but unsubdued. And the storm, as if exhausted in its battle with this man of iron will, or desirous of rewarding him for his heroic struggle, abated.

Once more the angry skies cleared away, the sun smiled brightly, and fair winds played among the rent sails, while the lashing waters melted into the long, regular, smooth-rolling swells which were more suited to the name later bestowed upon this fabled body of liquid. [*Pacific* means "peaceful" or "calm."]

Soon Drake found himself in the midst of the islands of Tierra del Fuego, and then, on a late day in October, he knew that he was really one of the great discoverers of the world.

He stood triumphantly on the southernmost point of land of the western hemisphere. [Check your globe or map!]

At his feet, where the dream of ages had woven a mystic shroud of romantic separation between them, the immense waters of the known Atlantic and the unknown Pacific rolled together in one mighty confluence of twin love.

As he traveled, Spanish ships were encountered, and prize after prize fell into his hands.

In Valparaiso harbor, the *Golden Hind* met a galleon heavily laden with Spanish plate to the amount of thirty-seven thousand ducats.

Portrait of Sir Francis Drake, his hand symbolically resting upon the world he circumnavigated.

Never before had a foreign sail been seen in those waters, and the crew of this vessel—the *Grand Captain of the South*—thought the new arrival must be a friend. They brought out bottles of wine to drink to the health of the Englishmen, but drank too deeply. Tumbled into the hatches of their own ship by their guests, these Spaniards did not succeed in extricating themselves until all their treasure had been carried away.

This was typical of many such adventures in the days that followed. To Drake and his men, the new ocean was anything but Pacific!

Then, one fine day, Drake pointed the bow of the *Golden Hind* straight across the Pacific. Past the Carolines, the Philippines, and the Moluccas, he made his way. Creeping among the maze of dangerous shoals and coral reefs in the Sea of Celebes, the vessel ran unexpectedly upon a sunken rock.

For twenty hours, she lay at the mercy of the waves, caught fast, resisting all efforts of her crew to release her.

Finally, as a last resort, eight guns and three tons of cloves were thrown overboard in an effort to lighten her sufficiently to float off. Fortunately, the wind freshened just then, and with her sails set to catch every ounce of pressure, she slid from the reef into deep water. The men uttered a heartfelt prayer of thanksgiving.

This was the last and greatest danger. Soon the *Golden Hind* cleared the Archipelago, rounded the Cape of Good Hope, and just two years and ten months from the day she sailed out of Plymouth Harbor, she again swept into it.

The long lost had returned.

In the meanwhile, England had given up ever seeing Drake and his crew again.

If he had not actually fallen victim to the fury of the big new waters, it had by this time become a foregone conclusion that he had either perished at the hands of some mysterious wild tribe along the coastline or in the islands or been taken and executed by the Spaniards.

Therefore, his return, laden down with spoils of vast richness, created a thunderbolt of surprise among his countrymen.

Indeed, Drake had arrived to find that the Queen had publicly disowned him—doubtless for political effect more than anything else. Even at the time, the Spanish ambassador was calling loudly for redress for the recent depredations that he had committed. ["Depredations" is often used to describe the destructions or robberies committed by a military force.]

Getting wind of this unpromising condition of affairs, Drake considered it prudent to drop anchor behind St. Nicholas Island, where his father had fled with him from religious persecution thirty years before.

From there, he dispatched a messenger to the Queen, with notification of his arrival, and was presently asked to appear before her august person.

When he obeyed the summons, it was not without full hands. Drake knew Elizabeth's fondness for money and jewels, and the richest of his spoils went with him as a gift to her and her courtiers.

The Queen, true to Drake's expectations, greeted him warmly. Pleased by the great value of these presents, and filled with admiration for her subject's daring exploits, Her Majesty loaded him with honors.

Already the hero of the hour with the countryside, his name was now, with the Queen's own commendations, on everyone's tongue. Throughout the breadth of England he was praised. His deeds were recounted before humble and pretentious firesides alike.

The bounty was carried to the Tower of London. But before registration, four hundred thousand dollars were extracted from the pile as Drake's own share of the spoils. Later, Elizabeth allowed him an additional fifty thousand dollars as a reward.

The *Golden Hind* was brought up the Thames River and there preserved as a memorial. On her deck, a public banquet was given, and the Queen, who graced the occasion, conferred upon Drake the honor of knighthood. He was now "Sir Francis Drake."

But the great seaman had not come home to be petted and pampered at court. His character was made of sterner and bigger things than that. Even now his active and far-reaching mind was burdening itself with thinking out other points for his attack.

However, five years were to pass before Drake could obtain his letters of marque [an official commission that allowed privately owned ships to capture merchant ships].

In the midst of treachery, vacillations, and other delays, his spirit fretted to be loosed once more upon the waters.

Meanwhile, he was forced to work at home in organizing the navy, voting supplies as a member of Parliament, and in improving the town and harbor of Plymouth, where he had been appointed mayor.

It was King Philip of Spain himself who precipitated matters finally.

The arrogant sovereign seized a number of British corn ships.

Elizabeth's temper was sufficiently inflamed for her to order Drake out to retaliate.

His fleet, quickly collected, was the largest he had ever commanded, and the most extensive privateering squadron on record up to that time.

It numbered two men-of-war, eighteen cruisers, and many pinnaces and store-ships, manned by two thousand soldiers and sailors.

On a late day in September 1585, Drake hoisted the English colors over his flagship, the *Elizabeth Bonaventura*, and led his fleet out to sea.

Again his goal was the West Indies, but this time, he was destined to strike a far more telling blow at enemy interests than before.

In his eagerness to get away, Drake began short on provisions. To replenish his stores, he stopped at the Bayona Islands and here helped himself to sufficient Spanish food to properly feed the entire fleet.

Arriving at St. Iago, of the Cape Verde group, he stormed the town and raided the island, then headed for San Domingo, which was reached shortly after Christmas.

San Domingo was a walled and strongly fortified city, the largest and most important in the West Indies. Drake realized that its fall would have a powerful moral effect upon the whole of Europe, and straightway made up his mind to attack it, even though this step would entail a serious naval operation for him.

He planned the assault with great care, surprised the garrison, and after a few hours' brisk fighting, the Spaniards fled across the river, leaving the storming party in possession of the Plaza.

Observing that his force was too weak to successfully defend the place for any time, Drake demanded a ransom of a quarter million dollars. He then destroyed the enemy's shipping in the harbor.

Cartagena, the capital, and one of the wealthiest of the many wealthy Spanish cities, was sighted in February. Formidable defenses surrounded it on all sides. From the sea, it was protected by a lagoon with access only by two narrow entrances, both well-guarded. From the land, any approach was made difficult by a creek, also amply fortified.

To storm the city seemed sheer madness, even for a force the size of Drake's.

But Drake, as usual, was resourceful enough to uncover a promising method of accomplishing his ends.

A detachment was ordered to wade through the surf and to come unexpectedly upon the city from a point where the enemy had made no provision for defense. At the same time, a boat attack was feigned on the side of the harbor in order to deceive the Spaniards.

The stratagem was successful. The city was taken with a rush; the defenders ran. Disdaining to loot the place of its rich treasure, Drake merely demanded a ransom of one hundred and fifty thousand dollars, which was gladly paid. Having also destroyed the shipping here, the commander then turned his eyes toward Panama.

But on the way thither, sickness broke out among his men. Many died, and others were stricken down every day.

With his diminished force, he saw that he could scarcely hope for success at so strongly a fortified place. So, in the latter days of March, he set sail for home, quite satisfied that the blows he had struck would do much to deter Spain in her threatened attack upon England.

The Defeat of the Spanish Armada

Defeat of the Spanish Armada, painted by de Loutherberg in 1796. On the left, a boatload of English sailors fight against the bow of a Spanish ship in the midst of flame and smoke.

PHILIP OF SPAIN, aided and abetted by the Catholic policy of the Pope and the court of France, had been making vast preparations to produce and equip the most powerful fleet the world had ever seen.

In all the ports of Sicily, Italy, France, Spain, and Portugal, vessels of enormous size were being built for him. The heaviest of guns were placed upon them, and vast quantities of provisions and small arms were collected.

England was kept in ignorance of Philip's intentions, rumors having been cleverly spread that the great Armada was designed to proceed to the Indies to realize extensive projects of distant conquest.

But England was not wholly deceived; the fact is, some of Elizabeth's advisors, prominent among whom were Sir Francis

Drake, had whispered in her ear what to look out for, and it was largely owing to her suspicion that her own country might be the object of all this enemy energy that she had consented to the last departure of Drake for the West Indies. Really it had been a most opportune blow that the famous sea king had dealt the Spaniards upon this occasion. But for the delay it caused in Philip's plans, the huge Armada would have been cast upon England much sooner than it came, and perhaps at a time when it could not have been successfully met.

On the afternoon of July 19th, 1588, a group of English captains were collected at the Bowling Green on the Hoe at Plymouth. [The Hoe is a large public space and promenade that overlooks Plymouth Sound, offering panoramic views of the sea. A "bowling green" traditionally refers to a flat, grassy area where the game of lawn bowls is played.]

Never before or since have men of such fame gathered at that favorite mustering place for British seamen. There was Sir Francis Drake, first English circumnavigator of the globe, the terror of every Spanish coast in the Old World and the New; there was Sir John Hawkins, the rough veteran of many a daring voyage on the African and American seas, and of many a desperate battle; there was Sir Martin Frobisher, earliest of explorers of the Arctic seas in search of the elusive Northwest Passage; there was Lord Howard of Effingham, High-Admiral of England, in his patriotic zeal bold even to disobey the Queen when obedience meant harm to the welfare of his country; and last, but in no wise least, you could also observe Sir Walter Raleigh, who had been commissioned to raise and equip the land forces of Cornwall, and who had come to consult with the Lord-Admiral and other high officers.

Many other brave and skillful mariners, though of less renown, were there also. All were enjoying a game of bowls on the green with true sailor-like merriment during their temporary relaxation from duty.

Presently Drake chanced to look off over the harbor. He noticed a small armed vessel running before the wind, all sails filled, making in furiously.

"What is this?" cried Drake, directing the attention of his nearest companions to the approaching vessel.

"Truth, and I believe 'tis none less I see than Fleming and his privateer coming yonder. A penny to know what could induce such mad speed from the lazy Scotchman!"

They were soon to know.

A few minutes later, the lanky Fleming stood in the midst of the excited group, telling the English officers that he had seen the great Spanish Armada off the Cornish coast that very morning!

Meanwhile messengers and signals had been dispatched fast and far throughout England, to warn each town and village that the enemy had come at last.

In every seaport, there was instant making ready by land and by sea; in every shire and every city, there was instant mustering of horse and man.

But England's best defense then, as ever, was her fleet.

Despite Elizabeth being slow in dispensing ammunition, doling out powder and cannon ball day by day until the last moment of need, Drake had exercised his men well at target practice, though even this the queen had regarded an extravagant waste.

It was a solemn sight when these two fleets had their first meeting.

The English ships, some of which had been loaned by the Dutch, totaled one hundred and fifty-two, manned by seventeen thousand men. The great majority of these were merchantmen, only thirty-six belonging to the Royal Navy. Practically all were light, swift, and easily managed.

The Spanish Armada itself was a gorgeous display, more fitted for a pageant than a war.

The great fleet consisted of one hundred and fifty ships, mostly much larger than those of their adversary. Of these, there were galleons, galleasses [large, heavily armed warships with strong hulls and numerous oars, allowing them to be both sailed and rowed], galleys [long, narrow ships propelled primarily by oars rather than sails], caravels [small- to medium-sized vessels with a distinctive hull design that allowed for better maneuverability and oceanic navigation. Caravels were used by explorers such as Christopher Columbus because they could sail long distances across open seas], petaches [small and nimble vessels], and zabras [three-masted ships that were known for speed and armed with cannons].

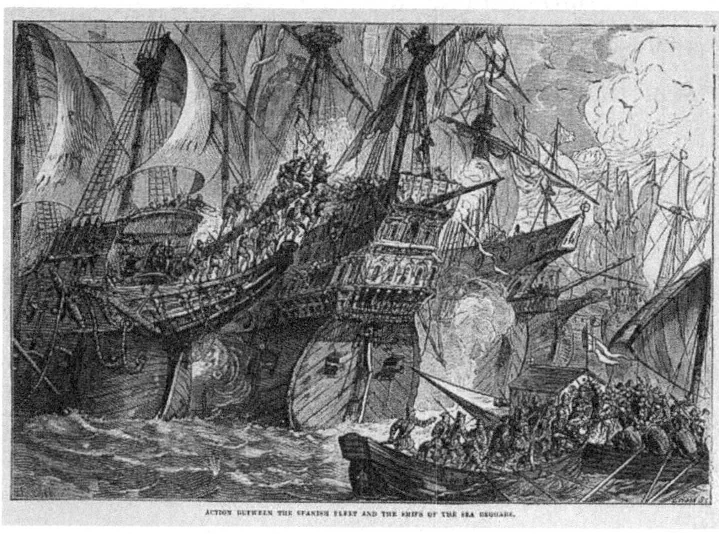

Fearsome Spanish galleons fighting at sea.

In a document of the time, Spain noted: "The number of mariners were about eight thousands; of slaves, twenty hundred and eighty-eight; of soldiers, twenty thousands (besides noblemen and gentlemen voluntaries)."

The document added that "the aforesaid ships were of an huge and incredible capacitie and receipt: for the whole fleet was large enough to contain the burthen of sixty thousands tunnes."

Interesting, too, is the following description in the quaint English writing of the period:

"The galeons were sixty-four in number, being of an huge bignesse, and very flatly built, being of marvelous force also, and so high that they resembled great castles, most fit to defend themselves and to withstand an assault, but in giving encounter farr inferiour are they unto the English and Dutch ships, which can with great dexterite weild and turne themselves at all assayes. The upperworke of the said galeons was of thicknesse and strength sufficient to bear off musket shot. The lower worke and the timbers thereof were out of measure strong, being framed of plankes and rigs foure or five foote in thicknesse, insomuch that no bullets might go through them, but afterward were found to sticke fast within the massie substance of those thicke plankes. Great and well pitched cables were twined about the masts of their shippes, to strengthen them against the battery of shot."

This great Armada was drawn up before the British in the form of a gigantic crescent whose horns were fully seven miles apart.

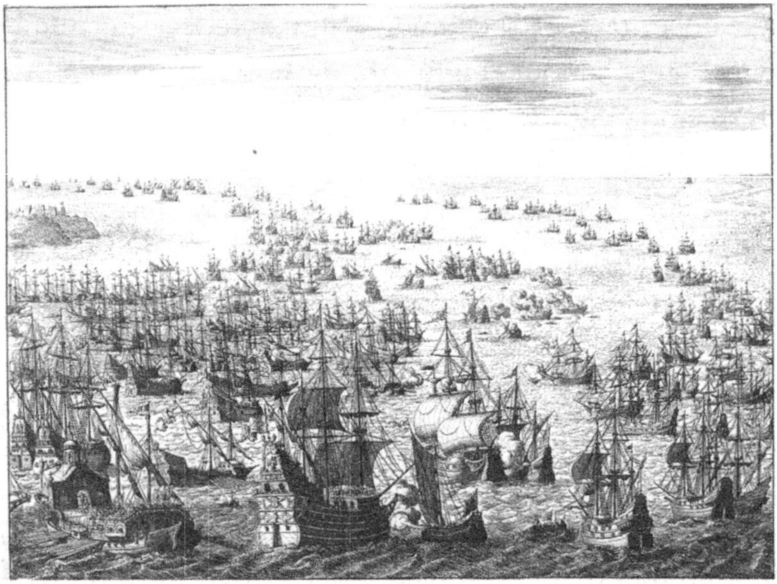

The Spanish Armada and the English lined up for the battle.

There was a southwest wind, and before it the huge vessels comprising the enemy fleet came sailing grandly on.

The English craft kept out of sight and range very neatly and skillfully, side-stepping and slipping around to the rear of the approaching enemy. Coming up swiftly from behind, they opened the attack.

Caught at a disadvantage at the outset, her soldiers far better fighters on land than on sea, Spain's huge fleet was dealt a stunning and staggering blow by the agile enemy whose guns were all manned by veteran seamen used to hitting their targets whether inanimate or animate.

After standing to for a brief period in an effort to stem the furious tide of shot which tore through their rigging, swept their decks and rained against their hulls, the Spaniards sailed away.

A running fight now took place in which some of the best ships of the Spaniards were captured, and many more were severely damaged.

For days the battle continued, always with the enemy on the go. The swift craft of the English sailed in and out and round and round among the unwieldy galleasses, cannonading them, and then escaping nimbly out of range, suffering little themselves.

Time after time they repeated these performances, teasing and harassing the clumsy Spanish boats and pelting their enormous turrets, which looked like castellated fortresses.

Undoubtedly had their crews been more equal, and the adversary not outnumbered in this respect more than two to one, the British lads would have closed with the Armada and fought matters out hand-to-hand. As it was, Lord Howard and Sir Francis Drake were wise to bide their time and to first ensure success by gradually weakening the adversary at long range.

Slowly holding their course along the coast, the two fleets, still fighting fitfully and bitterly, continued hostilities. Each day added not only to the spirit but to the number of the British force. Raleigh, Oxford, Cumberland, and Sheffield joined them.

The Spanish admiral also showed great judgment, and no mean skill, in his maneuvers with the awkward vessels under his command. Had he been any less of the clever navigator, he never could have brought himself and fleet through as well as he did.

And on the 27th of July he brought his Armada, sorely distressed but comparatively unbroken, to anchor in Calais Roads. Here, according to pre-arrangement, he expected to be joined by his countryman, the Prince of Parma, who was to come by water from Dunkirk with a flotilla and a large army with the purpose of invading England.

But Parma had been nicely held in the meantime by the Dutch and English blockaders, and so was still far away when the Spanish Armada entered Calais Roads.

The great Armada lay in the offing, with its largest craft ranged outside, "like strong castles, fearing no assault; the lesser placed in the middleward."

Lord Howard could not attack them in this position without great disadvantage. However, on the second night, past midnight, as the clouds covered the moon and no eye could pierce the darkness, eight vessels crept noiselessly within the Spanish lines.

A moment later the sea was suddenly illumined, and eight seeming volcanoes, spitting sparks and shooting long tongues of flame, bore down upon the terrified enemy. These were the dreaded fireships—such as the Greeks had so often used against the Turkish fleets in their war of independence—and had been unloosed by the forces of Howard and Drake.

As those eight masses of flame came nearer, a dreadful panic seized the Spaniards. The terror spread from ship to ship with seeming simultaneous quickness.

Amid yells of fright and the greatest of confusion, the Spanish crews cut their cables to avoid the oncoming conflagrations and took to the wildest flight.

One of the larger of the galleasses ran afoul of another, and both were so hopelessly tangled in a moment that they fell easy victims to their adversaries.

When daylight broke it was seen that the rest of the Armada was scattered along the Flemish coast. With great difficulty they obeyed their admiral's order to range themselves around him near Gravelines.

Now was the golden opportunity for the English to assail the crippled Spaniards once more.

Drake and Fenner were the first British commanders to reach the unwieldy leviathans and open the assault. Then came Fenton, Southwell, Burton, Cross, Raynor, followed by the Lord-Admiral himself and Lord Sheffield.

Huddling as close together as possible for mutual protection, the Spanish ships, firing desperately, broke and ran for it. After them, almost every shot on craft or crew came the relentless will o' the wisp opponents.

The towering ships of the compact Armada made capital targets for the British gunners—targets hard to miss within fair range.

Riddled, shattered, disabled, their own shots going wild or falling short, the best of the Spanish vessels soon gave up using their cannon, and drifted helplessly with the current toward the coast of Holland, past Dunkirk—and far away from the Prince of Parma, who, in watching their defeat from the coast, must have, as Drake afterward expressed it, "chafed like a bear robbed of her whelps."

And the remnant of Spain's colossal Armada—the sad, bedraggled little remnant left—made their way painfully through storm and hunger and sickness, to the shores of their mother country, still chased by the hound-ships of the British.

As the latter pursued, pitiful wreck after pitiful wreck of what was but a short time previous a floating armed palace filled with gayly dressed and elaborately fed Spanish noblemen and soldiery, drifted upon the tossing waves.

Of this great collection of proud craft only a pathetic handful ever returned to King Philip.

Much of the glory, if not indeed the greater share, of the defeat of the Spanish Armada was due to Drake.

Although occupying the secondary command of Vice-Admiral in the Queen's fleet, Drake had been at the forefront throughout the fighting and had been entrusted with executive powers that aided materially in the successful outcome of the battle. While his own vessel had been struck as many as forty times, no ship of consequence under his command had been seriously damaged; likewise, only minor casualties had been suffered among the crews, and no officers were injured.

The last act of Drake's tragic life was set among the scenes of his youth and early triumphs.

Drawn irresistibly toward those islands in the Caribbean Sea that had witnessed his first exploits, he led his squadron one day to La Hacha, to Nombre de Dios, and then, in a sudden wave of the adventurous spirit of his boyhood, he headed for Truxillo, the port of Honduras, and for the rich towns of Nicaragua.

But ill winds snared him; he was held in the fatal Mosquito Gulf, where pestilence lurked in every breath of the foul air.

Stricken with fever, his crew sailed with him back toward Puerto Bello.

There, on the 28th of January, 1596, the great sea king expired, and his remains were confined very fittingly to the sepulcher of the deep he had loved so much in life.

CHAPTER II

MARSHAL ANNE-HILARION DE TOURVILLE

Seventeenth-century portrait of Admiral Anne-Hilarion de Tourville

For by the dewy moonlight still,
He fed the weary-turning mill,
Or bent him in the chill morass
To pluck the long and tangled grass,
And hear above his scar-worn back
The heavy slave whip's frequent crack!

—*Whittier.*

A Knight of Malta

THERE WAS A time in the history of the chief maritime nations of Europe when it fell to the lot of each to seize and hold the supreme sovereignty of the waters.

Sometimes, this mastery was for but a very brief duration; sometimes, it was jealously maintained for years. But always was it kept at the cost of the greatest and most heroic effort, and lost because another power exceeded this endeavor.

One of these brilliant but short flashes of triumph came to France during the reign of Louis XIV, at the very outset of France's career as a recognized naval power.

Of the four illustrious seamen who were largely responsible for France's strength on the deep blue waters, none contributed quite so strongly as Admiral Anne-Hilarion de Tourville.

This valiant seaman was born at the castle of Tourville in Normandy in the year 1642, coming from a long line of noble and distinguished ancestry. His father, the Baron de Tourville, died when he was five years old. Slender and pale, almost delicate, it was little thought that this frail boy was destined to spend forty-five years of his life in an active and tireless service on the sea, and that he was to be counted the foremost commander of his time.

Tourville's noble birth automatically entitled him to become a member of the famous Order of the Knights of Malta.

As a preface to this distinction, at fourteen, he became a page to the grand master of the Order. He served as page for three years, was then placed on probation for twelve months for the higher honor of Knight, and when eighteen years old, stepped into the much-coveted position.

This was a proud day for the youth—a day that opened his long career as a seaman. The ensuing seven years of his life were spent on the Mediterranean, far from home, fighting the

Moorish buccaneers who swarmed over the narrow seas, and protecting the commerce of France from their ravages.

These wild corsairs [state-sanctioned pirates, authorized by their governments to attack and capture enemy ships], were the highwaymen of the Mediterranean, whose haunts lay along the creeks and inlets of the North African coast.

For more than one hundred and fifty years, these pirates had been stabbing the commerce of all of Christian Europe. They had spread terror along the southern shores of that continent; they had even ravaged the seaport towns of Spain, Italy, and Sicily; they had interrupted trading, held up rich convoys bound to distant marts, and chained thousands of unlucky Christian slaves to the galley benches of their robber craft.

Many of the greatest seamen of recent ages, admirals of Italy, Spain, France, and Holland, had spent the best years of their lives in hunting down and trying to annihilate these leeches of trade.

1600s painting: *A Sea Fight with Barbary Corsairs*. See the crescent flags?

But the pirates' natural rivals and untiring enemies, those whose special mission it had become to dispute and weaken their power, were the Knights of Malta, themselves the Christian pirates of the Mediterranean, whose bounty was these corsairs from the Barbary steppes.

The brilliant crusades against the Moorish crescent were carried on in a fleet of galleys most splendidly equipped.

Every Knight was obliged to serve in four cruises of six months each, and most were not sorry when their period of stirring adventure came to an official end.

Young Tourville particularly started out upon his voyages with gusto. In his very first encounter with the Barbary pirates he exposed himself heedlessly to the sweeping fire of the enemy, and fought with such abandon of bravery that he won the admiration of friend and foe alike. In this encounter he was wounded in three places but insisted on staying on deck and using his gun.

Then came a long succession of heroic deeds. These brought him a reputation for ability and intrepid courage that spread from Venice to the royal court of France.

The Venetian republic was so grateful for his services in relieving her from the depredations [Remember that definition?] of the Algerian sea-rovers that she bestowed upon him the titles of "Protector of Maritime Commerce" and "Invincible Seaman."

As a sequel, Louis XIV, his own ruler, sent for him and personally complimented him upon his fine exploits. Shortly afterward he received his commission as captain in the Royal Navy and was given the command of a ship.

Portrait of King Louis XIV of France by Claude Lefèbvre, 1670.

Notice the ships in the distance behind the king of France.

Meanwhile the relations of France to other continental nations became strained.

Finally, in 1672, she and Great Britain united with the purpose of crushing the power of the Netherlands, as stated in a preceding section of this volume.

In this step, France was largely motivated by her feeling of security in the navy which she had recently built up—a navy such as she had never owned before, a navy that was the greatest pride of the King's heart, and one which he was very desirous of trying out, like a boy with a new kite.

Envying the enormous commercial wealth of the Netherlands, with Louis XIV at least, was a secondary consideration.

Great Britain, weary of hostilities, had withdrawn from the contest and had signed a treaty of peace with the Dutch. France was left to continue the war alone.

Determined to destroy the commerce of the Netherlands, she sent her men-of-war to the Mediterranean.

In the battles that followed in southern waters, Tourville took an active part under the leadership of Admiral Duquesne.

With him were many other gallant young Frenchmen, the pick of all France, but of them all none stood out so brilliantly in action as the dashing, slender figure of Tourville himself, then thirty-four years old.

Allied with Holland, Spain was threatened with the loss of one of her most important possessions in the Mediterranean. This was Messina, a seaport in Sicily, the queen island of the South Sea. The people had risen in revolt against the Catholic King, captured the forts, and gained almost the entire possession of the city. Too weak, however, to maintain the rebellion without foreign aid, the Messinese asked for the protection of the King of France.

At this stage of affairs, Admiral Vivonne was lying at anchor with his fleet off the coast of Catalonia. With him were such renowned seamen of France as Valbelle, Preuilly, and Tourville. He dispatched Valbelle on September 27th, 1674, with a small squadron to help the insurgents. This failing to meet the demands of the occasion, in January a second expedition was sent with Tourville as captain.

Several forts had been retaken by the Spaniards in the meantime, among which were the Pharo and Reggio. Outside the gates of Messina was encamped a Spanish army. This force, aided by a fleet of forty-one sails that guarded the entrance to the Straits, pressed the city vigorously.

Messina lay surrounded by land and sea, practically at the mercy of the Spaniards when, on the 2nd of January, Valbelle appeared with his small squadron consisting of six warships, one frigate, three fireships, and a convoy of supply vessels.

Portrait of Jean-Baptiste Valbelle, French naval officer, known for his role as commander of a squadron of French ships during the Franco-Dutch War of 1672–78. Artist unknown

Favored by fresh winds and an incoming tide, the French flotilla dashed through the channel, swept past the barricade of Spanish galleys, weathered the fire of the forts, and forced an entrance into the Straits with bewildering audacity. Amazed to inaction, the heavy men-of-war offered little resistance. Messina was relieved by a brilliant stroke.

But the new supply of provisions could not last forever. At the end of five weeks, the stores were exhausted. Starvation faced the rescued. The sorely besieged prayed that help might reach them before it was too late.

That prayer was answered.

In February Tourville's ships sailed into view, and soon followed more French ships in the shape of the squadrons of Duquesne and Vivonne and Valbelle.

This great force was more than the Spaniards could withstand. In a short time, they were put to flight, leaving half their number behind. Needless to say, the victors were received with the wildest joy and greatest gratitude when they entered Messina and brought huge stores of food to the besieged rebels.

During the remainder of the year of 1675, the French strengthened themselves in Sicily, and extended their conquests along the southern coast.

On the 17th of August twenty-nine ships under Vivonne anchored in the bay before Agosta and opened fire on the forts protecting the harbor.

While this was going on, Tourville was leading six ships to the mouth of the harbor. Despite the galling fire to which he was soon subjected, he forced an entrance and turned his broadsides upon Fort Avalo, the strongest and most important of the enemy defenses.

As the fort still held out after a brisk cannonading, Tourville sent Captain Cöetlogon with a small landing party to make an attack at closer quarters.

Under a storm of shots and stones, the first barricade was then taken, but Cöetlogon's little force was threatened with capture by his numerous foes. Alarmed for the safety of his friend, Tourville threw himself into a small boat with a few volunteers and flew to the rescue.

The Spaniards made a feint by running up a white flag. When the rescuers had come within range, they let loose a wild storm of artillery in their direction.

But Tourville was not to be nonplussed. After an hour's obstinate attack, the second barricade was carried, the fort surrendered, and the first landing party was saved. The town quickly thereafter capitulated.

In reporting this success, Tourville, as customary, gave the lion's share of the credit to his officers and men. But his government, while up to this time slow to promote him, was not blind to the truth, and a year later the modest and intrepid captain was rewarded with a commission as commodore.

Wooden statue of Tourville from a maritime museum in France.

Punishing the Barbary Pirates

IMAGINE YOURSELF GAZING down upon the volcanic Islands of Lipari—the very outposts of Sicily—the rock sentinels that guard the entrance and act as finger posts for many a grizzled mariner as he feels his way along the shores of Italy past her beautiful Queen Island.

On Stromboli, which forms the gateway to the gulf, stands the great lighthouse of the southern sea.

Rising more than two thousand feet above the blue waters into which it has so often spat its molten saliva, Mount Etna in solitary grandeur furnishes by night the ever-burning beacon to seamen which is denied to Lipari.

At the foot of this centuries-old volcano, under the open maw of fiercely boiling flame, turbulent gases, and gurgling lava, the greatest sea fighters of the age have now gathered to measure their skill and prowess.

Now, sent by William of Orange to the Mediterranean to confront the Spanish fleet and force the rebel Sicilian town to return to the allegiance of the King of Spain, the famous De Ruyter faces the French fleet with eighteen Dutch men-of-war. Duquesne's force numbers twenty-five warships and six fireships. He has divided his flotilla into three squadrons.

One squadron is commanded by Tourville.

For twenty-four hours the rival fleets lay watching each other warily, maneuvering to gain advantage of the wind.

As the first gray light of dawn begins to appear in the east on the morning of the 8th of January, 1676, a stiff breeze springs up to the advantage of the French.

Without hesitation, Duquesne orders sails set and bears down upon his enemy.

Warships in battle, circa 1600s

The Dutch receive the onslaught with a firm front, retaliating with heavy fire. From ten in the morning till ten at night both sides fight with vigor, often coming to close quarters.

From the deck of his ship, the *Sceptre,* Tourville supports his leader valiantly, never failing to be in the hottest part of the combat. On one occasion he is violently attacked by a huge three-decker. The latter's heavy guns soon create havoc with his rigging, and his ship, crippled and shot torn from bow to stern, is only saved by a French fireship that comes to his assistance.

Just before nightfall, a squadron of nine Spanish galleys pounce down upon the French flagship and annoy her with their powerful chase guns, but Tourville sends several thirty-six-pound cannon balls among them, and they scatter.

When darkness comes, the loss to the combatants has been about equal, each suffering considerably.

The next day, both fleets receive reinforcements.

But the battle is not renewed; Duquesne sails around Sicily and enters Messina harbor from the south; De Ruyter goes first to Naples and afterward to Palermo to resupply and refit.

The most important action, however, was yet to take place.

On April 22nd, between Catania and Agosta, the French fleet of thirty men-of-war encountered the combined squadrons of Spain and Holland, numbering three fewer ships.

At first, bearing down quietly upon one another, there was no sign of hostility—until they were within musket-shot range. Then suddenly both fleets let loose violent broadsides.

The conflict was prolonged and furious. Toward the close of the day De Ruyter's flagship succeeded in closing with the *Saint Esprit*, at whose masthead flew the ensign of Duquesne.

In a moment, the two rival flagships were doing their utmost to sink the other. All their available guns were aimed with the deadliest intent, and men stood ready with boarding tools and weapons for the time when the crucial moment should arrive.

At this juncture Tourville observed his chieftain's predicament.

With the *Sceptre* and *Saint Michel*, he appeared quickly on the scene, with a ship at either side of the Dutch flagship.

Neatly trapped, it was by the utmost difficulty and only with a fine example of his famous seamanship that De Ruyter managed to extricate himself without capture.

However, he had been sorely wounded himself. Not until later did this knowledge come to the French—not until their brave adversary's fleet had retired to Syracuse and De Ruyter's dead body was consigned to the sea by his sorrowing countrymen.

By ten o'clock that evening the combat ceased.

Next morning, in rain and mist, the combatants separated, the French to keep to the seas until May 1st, after which they sailed into the harbor of Messina.

The death of De Ruyter was an irreparable blow to the Allied fleet. Robbed of the strength of its most able commander, it is no wonder that the next sea battle fought by the French proved to be one of the most signal naval triumphs on record.

After restocking and repairing their ships, the Allies sailed out of Syracuse harbor, doubled the island of Sicily, and entered the port of Palermo, where they intended to await the movements of the French fleet. Their idleness was short lived.

Dutch man-of-war ship (to the right) approaches a harbor on the Mediterranean Sea.

On May 28th, the French fleet of twenty-nine warships, nearly as many galleys, and almost a dozen fireships were put to sea from the harbor of Messina. The fleet passed through the channel and sailed northward on a lookout for the enemy.

The Duke of Vivonne, viceroy of Sicily and nominal head of the Mediterranean fleet, was a member of the expedition, having come along in person that he might share what was considered to be the last glorious venture of French arms. His flag was run up on Tourville's ship, the *Sceptre*, which thus became the chief vessel of the center.

Something like four days after leaving Messina, the French fleet sighted Palermo. There, before their eyes, the complete fleet of the Allies rode at anchor, arranged in battle formation.

On the right and left wings were the Dutch ships under the command of Admiral Haan, who had succeeded De Ruyter. In the center, the Spanish vessels were concentrated under De Ibarra. The craft were three or four cable lengths from the entrance. In all there were in the combined Dutch-Spanish fleet twenty-seven ships-of-war, nineteen galleys, and four fireships.

No sooner had the French sighted the enemy than Vivonne called four of his most trusted officers for a most difficult and perilous undertaking. Among them was Tourville, the youngest of the commodores. These officers were asked to make a complete examination of the adversary's position, and to draw up a plan of his defenses.

The party set out in a small sailboat in broad daylight. Supported by the squadron of galleys, which were ready to protect them with their guns at the slightest provocation, the valiant and devoted quartet of young Frenchmen entered the harbor and approached within close range of the battlefront.

Sailing up and down the enemy's line, Tourville and his companions made careful observations, no detail of the arrangements of defense escaping their vigilant eyes. Struck with admiration at their audacious bravery, the Dutch and Spanish looked on silently. Not a gun did they fire.

Vivonne immediately called a council of war to decide on a plan of attack. Tourville had already worked out a plan of procedure in his active mind, which he now presented to his commander-in-chief. So pleased were the officers with this, especially Vivonne himself, that after some discussion, it was accepted with great enthusiasm.

A stiff breeze was blowing from the northwest the next morning, the 2nd of June. Shortly after daybreak, aided by this breeze, the French fleet sailed, in battle order, through the entrance to the harbor. They were led by nine selected warships and five fireships which were to open action by attacking the head of the enemy's line.

Every deck was cleared for action, every man was at his post.

But before the French had swung into place, the Dutch opened fire.

Their heaviest broadsides were poured into the bold intruders. In return, the French sent even more furious and well-directed charges across the waters at the Dutch. This made their ships quiver and waver.

Along the whole line, the French fire was fully as terrible. Coupled with the deadliness of their aim, the Allies were soon filled with fear and dismay. In less than half an hour, the Spanish vice-admiral had cut his own cables and drifted toward shore in a spasmodic effort to save his ship from capture.

With the line open, the French renewed their attack with greater vigor than before. Two more flagships were compelled to cut their cables, and in the resulting panic, other Dutch and Spanish ships followed suit. The French made good use of their fireships, through their agency burning twelve enemy men-of-war, among them a Spanish and Dutch flagship. The *Capitane* and *Steenberg*—Spanish and Dutch vessels, respectively—blew up with fearful force, covering the surrounding bay and some of their surviving craft with burning debris.

Dutch ships at war, circa 1650s.

As the flames gathered headway and communicated to neighboring sister ships, worry seized the Allies. Wild with terror, Dutch and Spaniards alike fled for refuge. Here, under the burning cannon balls and shot-spraying grenades falling gin showers on the city of Palermo, the fleet fled inland.

The victory could scarcely have been more brilliant and complete. By it, the French had asserted their mastery of the waters of the Mediterranean. The very blow that had relegated the strong navy of Spain into the dust also raised the navy of France to the pinnacle of sea powers.

The famous treaty of Nimègue [a Dutch city], signed on August 10, 1678, between Louis XIV on the one side and half of Europe on the other, marked the beginning of that dazzling period of naval supremacy that, for almost fifteen years, placed France for the first and only time in her history at the head of maritime nations.

The flag of every country saluted the standard that floated proudly from the mastheads of the ships of France.

Naval combat scene, drawn during the 1600s by an Italian draftsman.

Not long after this, action on the seas again became imperative for the French.

The commerce of Europe had been endangered by the piratical swoops of the pirates of Barbary.

In order to effectually put a stop to these bare-faced robberies of her merchantmen, France decided it was advisable to attack the miscreants in their main retreat—Algiers [North Africa].

To this hiding place, the pirates would lead their captured prizes, fortifying themselves against interference in its ample, well-caparisoned harbor.

In Algiers, it was truly suspected that thousands of Christian captives languished in foul prisons and in wretched servitude.

So in the summer of 1682, and again the following year, a French fleet under Duquesne and Tourville was sent to do what it could to clean out this nest of the pirates. For the first time new weapons were to be used. Among the heavy ships were small, flat-bottomed boats called bomb-galiots, each of which carried two mortars and four guns.

Example of a bomb galiot, anchored in a bay

Reaching the harbor of Algiers, the Frenchmen lost no time in letting the sea robbers know they meant business.

From the warships, cannon balls were thrown into the beautiful Moorish city, and from the mortars on the newly invented galiots, death-carrying, destructive bombs were rained upon the roofs of the buccaneers' homes and storehouses and places of amusement, as well as other buildings.

Palaces and mosques fell in a mass of ruins; storehouses were wrecked or burned; houses crumbled. The wives and children of the pirates, the pirates themselves, their slaves and captives, could be seen running through the streets in frantic quest of better shelter.

Many captives took advantage of the terror of their brutal captors to run down to the water's edge where, with outstretched arms, they mutely pleaded for rescue.

Tourville—now a vice-admiral, and the first in every perilous enterprise—came and went in a small boat, subjected to incessant fire from the shore, to direct and watch the work of the mortars. His heart ached for the begging captives, but it would have been suicide to have attempted their rescue just then.

Finally, the Algerians sued for peace [asked for a truce]. But Duquesne and Tourville refused to listen until every Christian captive had been delivered from bondage.

Then, for five days, there was silence on the bay and a respite in the city, while boatload after boatload of Christians were carried from the shore to the ships that were to take them from slavery back to their almost-forgotten homes.

Half-starved, hollow-cheeked, ragged, eyes dulled till only the fever-flame of life's last piteous grip gave them Divine expression at all, more than seven hundred white slaves of the Barbary pirates were thus restored to liberty.

It was a proud day for Duquesne and Tourville, for it had brought them a greater trophy than just mere victory over an enemy at sea. No quantity of prize warships could ever equal their satisfaction at this saving of their fellow men from such cruel bondage.

CHAPTER III

COMMODORE JOHN PAUL JONES

An 1830s portrait of John Paul Jones

The Stars shall glitter o'er the brave,
When death, careering on the gale,
Flag of the seas, on ocean's wave,
Sweeps darkly round the bellied sail,
And frightened waves rush wildly back
Before the broadside's reeling rack;
The dying wanderer of the sea
Shall look at once to Heaven and thee,
And smile to see thy splendors fly
In triumph o'er his closing eye.

—*Halleck.*

From Toy Ships to Big Ships

John Paul Jones was a man of no country, a "citizen of the world," a fighter on the side of humanity, the bitter enemy of the brutal oppressor. Likewise was he a Scottish trader, an American commodore, a French chevalier, a Russian admiral, the father of the United States Navy, the winner of the greatest sea battle of the Revolutionary War.

Fittingly has it been said of this striking sea man, "He was the Drake of the New World." Like Drake, he was a man of strong individuality and stunning contrasts, a dashing adventurer, a gracious courtier, a brilliant commander, a strict disciplinarian, a tender-hearted and loyal friend, a ferocious avenger, an invincible fighter. Unlike Drake, he held his spirit of plunder and flaming ambition under a tempering check that made him the more admirable.

Again, like Drake, his name has been woven with a romantic tradition and fanciful popular legend that has clouded and distorted actual truth and the records of history. And, like Drake, he has been as much hated in Great Britain as was *El Draque* in Spain.

John Paul was born on the 6th of July, 1747, at Arbigland, on the southern shores of Scotland. His father was a Lowlander, head gardener, and gamekeeper to a country squire, while his mother was a Highlander and a descendant of one of the fierce clans that had their homes among the heathered hills.

John Paul had three sisters and three brothers, he being the youngest. His oldest brother, William, was early adopted by a distant relative, William Jones, who had emigrated to the American colonies and lived on his plantation in Virginia. The adopted lad at once assumed the name of Jones. Little John

Paul himself had never seen this brother, as he had been born three years after William departed.

John Paul's childhood sped quickly; he soon grew into a hardy, active, self-reliant lad of twelve, at which time his scanty instruction in the parish school of Kirkbean ceased.

In the meanwhile, he had mastered other lessons that had set his adventure-loving young soul into a wild flame of restless yearning. These lessons came from his frequent trips to Carsethorn Creek, a nearby stream where fishermen were wont to seek shelter from storm and tide and unload their cargoes of tobacco for Dumfries.

In these magic waters, he had sailed his mimic boats; on the weather-worn wharves, he had listened with wide-open eyes to the stirring tales of the old Scottish sailors, and often watched with eager interest and keen intelligence the coming and going smacks of the fishermen as they steered their way through the narrow lanes of shipping.

On these waters, he had already learned to handle a yawl and to brave the northeast squalls that so often tried the courage and skill of many an older seaman.

Young John Paul now began to beg his father to let him go to sea. His passionate longing was gratified. The father, before his son had yet reached his thirteenth year, sent him across the Solway [an inlet that forms part of the border between England and Scotland] and apprenticed him to James Younger, a prosperous merchant in the American trade.

As master's-apprentice, John Paul was now a full-fledged sailor on board the *Friendship*. How proud he was!

A month later, after a stormy voyage, this vessel dropped anchor in the Rappahannock River in Virginia, not far from the plantation of William Jones—the relative who had adopted his eldest brother.

The meeting between the two brothers, neither of whom had ever seen the other, was a most touching one.

John Paul found William Paul Jones a successful married man, and overseer upon his adopted father's plantation.

While the *Friendship* lay at anchor, the young boy spent much time on land with his brother, and it was then that he was for the first time attracted to the novel and independent life of the American colonists—a life that had a peculiarly strong appeal for his own rugged spirit.

William Jones took a fancy to the lad, and offered to adopt him. But while his lively interest had been awakened in the half-civilized new country, John Paul's real love was for the sea. He therefore declined, and soon sailed away for home.

John Paul Jones, 1781

For the next six years, John Paul sailed on trading voyages in the ships of Mr. Younger.

In this time he advanced rapidly in sea knowledge and skill. He had a keen, open mind, and a retentive memory, and never thought he "knew it all."

With these characteristics, he was sure to gain the utmost information and to thereby pave the way for success in his future undertakings.

In 1764, he was serving as second-mate on West Indian traders; a year later he was promoted to first-mate. Then Mr. Younger retired from business, released him from his indentures, and rewarded his faithfulness with a sixth interest in a packet engaged in West Indian trade.

This packet was the *King George*.

With this vessel, John Paul made two voyages to the west coast of Africa after slaves. The best men of those days thought nothing wrong in such transactions, not having their minds awakened to the full horrors of the practice as have the men of today. So, if the conscience of John Paul pricked him a bit during these financially profitable trips, he kept the fact pretty much to himself and history never knew it.

At the end of his second voyage after slaves, our hero sold his share in the vessel to Captain Benbigh. At Kingston, Jamaica, he boarded the *John O'Gaunt* as a passenger, bound for Whitehaven, England.

Little did the young sailor know, or the crew, that when the trader sailed out of the Antilles and into the Atlantic Ocean she carried with her the dreadful germs of the yellow fever.

Barely had she cleared the Windward Islands before the ravages of the scourge began to spread among the crew. Within a week, the captain, his officers, and most of the deckhands had succumbed and their remains were consigned to the deep. Only six human beings were left on the ill-fated ship. One was John Paul, the passenger.

The young man's sea experience and sea study were now to stand him in good stead.

There being none left who were competent to handle the ship, John Paul took command. With neatness and dispatch he guided the fever-stricken brig across the dangerous waters of the Atlantic, and brought her safely into the harbor at Whitehaven.

So pleased were her owners, Currie, Beck & Company, that they gave him a generous monetary reward, in addition to which they appointed him captain and supercargo of a new ship—the *John*—which was engaged in making trips to the West Indies.

In command of this vessel, John Paul made three voyages. During the course of these he again visited his brother on the Rappahannock, and the bonds of affection for America were drawn still closer.

When William Jones died in 1760, he left his entire property of three thousand acres, buildings, slaves, cattle, and a sloop, to his adopted son.

The will provided that, should the adopted son die without children, the property was to go to his youngest brother—our John Paul.

When John Paul sailed away for England after his last visit to William, he had no thought that he was soon to return, the owner of this fine American plantation.

The ensuing two years were passed by John Paul in making voyages to the Indian Ocean as a convoy to transport ships of the East India Company.

1800s portrait of John Paul Jones, artist unknown

His last merchant trip was in late in 1772 on the vessel *Two Friends*.

Sailing by way of Lisbon, the Madeira Islands, and Tobago, he dropped anchor in the Rappahannock River in April of the following year. Expecting a hearty welcome from his brother, he was grief-stricken to find him lying at the point of death, quite unconscious.

At William's death, John Paul became master of the large estate in Virginia. He decided to assume the last name of Jones, to fit more gracefully into the title of his new possessions, and sent the *Two Friends* on her homeward voyage under the command of her first-mate, with word to her owners that he intended to become a Colonial planter in America.

For upwards of two years, he really enjoyed the quiet and independence of the new life that had opened up to him as from the tip of a fairy's wand. During this period, he left the active management of the estate to his brother's faithful overseer, and devoted the greater share of his own time to study and society.

He mastered French and Spanish, naval history and tactics, diplomacy, and politics. He entertained the neighboring families with lavish hospitality. He traveled extensively.

Thus the poor Scottish gardener's son, educated largely by his own exertions, became a scholarly and educated American of the day.

Moreover, he was soon, very soon, to make valuable use of his learning and skill in the cause of his new country.

The *Drake* and the *Ranger*

IN THE SPRING of 1775—that eventful spring when the war of the Revolution embroiled America and England—John Paul Jones was leisurely making his way in his sloop [a sailboat with a single mast] along the coast of New Jersey on a pleasure sail to Boston.

Upon reaching New York, he encountered William Livingston. [Livington would play a large part in the American Revolution. He was the first governor of New Jersey during the war.]

Livingston's face looked very serious. John Paul saw at once that something had gone wrong, and Livingston's first words proved him right.

"John, have you heard the ill news?" asked Livingston.

"I have not been favored," replied Jones. "I trust it is nothing serious concerning your own family?"

"I fear it *is* serious, but it concerns my family no more than all families in the Colonies. John, my friend, word has just come that the British have beaten us at Lexington!"

Instantly John Paul Jones was as concerned as his friend.

He plied him with many questions.

After a long discussion, they parted.

Jones immediately gave up his plans for going to Boston; he turned the bow of his sloop back homeward, and three days later, from his plantation, was writing to members of the Continental Congress and offering of his services.

In June the new marine committee invited him to present any information and advice he could on the selection of suitable naval officers and ships for the beginning of a navy to defend the American seaboard from the seacraft of England.

John Paul Jones's suggestions met with favor. As a consequence steps were at once taken for carrying them out with speed and dispatch.

The first squadron of America's navy consisted of four ships—the frigates *Alfred* and *Columbus*, and the brigantines *Andrea Doria* and *Cabot*.

The first list of officers embraced five captains, five first lieutenants, and eight junior lieutenants. In this list, John Paul Jones stood at the head of the first lieutenants. Favoritism among some of the powerful members of the marine committee kept him out of the list of captains, a position to which his high attainments surely qualified him.

Of course, Jones felt the injustice of it all. But in his selflessness and broad love of service for service's sake, he made no allusion to the slight except in a letter sent about that time to a friend. In it he says:

"I am here to serve the cause of human rights, not to promote the fortunes of John Paul Jones. . . . I will cheerfully enter upon the duties of first lieutenant of the *Alfred* under Captain Saltonstall. Time will make all things even."

And time did "make all things even," as we shall see.

Portrait of John Paul Jones, artist unknown. Notice the ships on fire behind him.

Not until the 17th of February was the little American squadron ready for sea. On that day the light, sleek ships sailed out of Delaware Bay and headed for the Bahama Islands.

Almost two months later, the squadron returned to home shores after a cruise that was productive of little more than a showing of disgraceful incompetence on the part of many of the officers.

As a result, captains appointed through influence and favoritism were dismissed, while Jones was honorably retained and given an independent command.

The *Providence,* his new ship, was a small sloop of fourteen guns. After using her to transport troops and stores between New London, Connecticut, and New York, and convoying American ships along the coast, he went out for six weeks to harass British commerce.

With a crew of seventy men, he sped through waters swarming with British frigates, from the Bermudas to Nova Scotia. He destroyed the enemy's fisheries at Canso, and made two daring descents on the island of Madame.

In all, sixteen prizes fell into his daring hands, besides a large number of fishing boats. Eight were manned, the remainder destroyed for want of sailors with which to sail them home.

Near the Bermudas he encountered the British ship *Solebay* with its twenty-eight nine-pounder guns. For six hours he was chased by the more powerful enemy, part of the time within range and once almost within the *Solebay*'s clutches.

But, by a very clever maneuver that surprised and caught the Britisher at a disadvantage, the American frigate succeeded in escaping.

Believing that he could be of much more service to the American cause in foreign waters than at home, John Paul began to appeal for permission to sail to the English and Irish channels.

Final recognition came from General Washington himself who said:

"Captain Jones, you have conceived the right project, and you are the right man to execute it."

The result was an appointment to command the new sloop-of-war *Ranger*, which carried 20 six-pounder guns. John Paul was told to hold himself in readiness for a swift sail to France to carry dispatches of the highest importance.

The *Ranger* sailed out to sea on November 1st, 1777. The news she carried was the surrender of General Burgoyne at the Battle of Saratoga.

Spreading every inch of canvas that he could, Jones forced his ship through the Atlantic at top speed.

By day the stanch little *Ranger* staggered on in the teeth of heavy northeast winds and blinding snow squalls; by night, beating her way through darkness and blanket fog, she continued just as bravely but with necessarily diminished speed.

The second-lieutenant states that the Commodore John Paul Jones declared he would "spread this news in France in thirty days."

The way the sloop bounded along surely looked as if he would keep his promise.

On the last day of the run, the *Ranger* captured two prizes. She sailed into the Loire River in France and dropped anchor at Nantes.

Commodore Jones hurried to Paris with his packet of news, only to find that he had been outstripped.

Mr. Austin, who had sailed from Boston two days earlier with duplicate messages, had arrived just twenty-four hours before.

But this was not the only disappointment to greet Jones in Europe.

He had been promised a large new frigate, built at a neutral Dutch dockyard for the United States. On reaching Paris he discovered that the vessel, for political reasons, had been sold to France.

So, instead of starting on his long-cherished cruise in British waters on the deck of a forty-six-gun frigate, he was forced to content himself with his little sloop of less than half that armament.

It was a dark outlook.

Nobody but a dauntless commander and an equally dauntless crew would ever have thought of fighting with such an insignificant craft. In fact, most captains would have turned back home, disgruntled and timorous.

Not so John Paul Jones.

With what he had he would do the best he could.

The Scotchman gave his ship a thorough refitting, and early in February, he sailed into the harbor of Brest, in France. A dense mass of ships rigging filled the view. Jones saw that it was the great French fleet under Count d'Orvilliers.

Bent upon upholding the honor of the new American flag, which was gayly flying from his masthead, the Scottish Commodore asked, as a condition of his entering the port, that a salute be given his colors. His request was granted. As the national emblem passed with dignity through the lanes of heavy battleships, the French guns roared out the first salute ever given by a foreign navy to the standard of the United States of America.

First Recognition of the American Flag by a Foreign Government, 14 February 1778. By artist Edward Moran, 1898, showing the Continental Navy Ship *Ranger*, commanded by Captain John Paul Jones, receiving the salute of a French squadron.

Just one week earlier, the Treaty of Alliance between France and America—which first recognized American independence—had been signed at Versailles, France.

The salute to the flag was a symbolic seal to the treaty.

To Jones, it was a matter of strong personal feeling. Not long before, when the same resolution of Congress that had appointed him to the command of the *Ranger* had also decreed that the national flag should be thirteen stripes, alternate red and white, and thirteen stars in a blue field, he had said, "The flag and I are twins."

This association was now strong upon him.

On the morning of April 24th, the *Ranger* was off Carrickfergus, on the north coast of Ireland. Inside the harbor, and just on the point of coming out, was the *Drake*, a British sloop-of-war of twenty guns. Contrary winds and an incoming tide had delayed her.

Now, as she saw the stranger in the offing, she sent out one of her boats to reconnoiter him.

Lured on by the innocent-looking stern of the *Ranger*, the British scout came a little too near. The first thing her crew knew, they were pounced on and made prisoners.

Apparently the *Drake* resolved to see why her men failed to come back.

Out she came just before sunset, approaching within hail in mid-channel. The British flag went fluttering up to her mast-head.

At the same time the Stars-and-Stripes was flung out to the breezes by the *Ranger*.

"Ahoy! what ship is that?" bawled the British captain.

It was an unnecessary inquiry—one fraught with the confusion of the moment. The answer was perhaps as equally unnecessary.

But it came:

"The American Continental ship *Ranger*! Come on; we are waiting for you!"

Scarcely had the last word of the reply died away when there came a raking broadside from the American at close range.

The *Drake* bore up, and poured back her own shot.

The engagement between the British ship-of-war Drake and the American Continental ship Ranger.—Page 93
Life of Paul Jones.

British ship of war *Drake* and the American *Ranger*.

For more than an hour, the firing continued, deluge after deluge of lead being thrown from one ship at the other.

But the aim of the British gunners was not as good as that of the Americans. Most of the enemy fire passed harmlessly overhead or fell short, while that of the men on the *Ranger* found some telling mark.

After a while, the *Drake* was in a sad plight.

Her spars and rigging hanging ragged and uselessness; her sails were fluttering and flapping remnants of riven canvas; her hull was polka-dotted with further evidence of the good marksmanship of her adversary. Her commander fell, killed by a ball. Here and there on her deck lay many of her crew, wounded and dead.

At last, when she had become "an unmanageable log on the water," she was boarded by the victors.

The capture of the *Drake* was the first brilliant naval success of the war.

The ships themselves were small and unimportant, it is true, but the results were large in comparison, and a great moral victory had been accomplished.

This triumphant cruise of the little hornet *Ranger* in waters far from home, her seizure of numbers of prizes, her bold fire-brandery in the harbor at Whitehaven, and her even bolder capture of His Majesty's ship *Drake* right under the noses of the British nation, not only formed a grand and dramatic début of the new-born navy of the United States, but it aroused and alarmed the enemy coast, filling all Englishmen with stunned surprise, indignation, and humility.

In addition, it gratified and won the admiration of France, the foe of Britain and friend of America.

And most important of all, it brought to the bitterly tried hearts of the American Colonists in the distant new republic new courage, new confidence, new strength, and new visions.

The *Bon Homme Richard* and the *Serapis*

WHEN HE REACHED Brest, France, with his prize, which was the first trophy of the war to reach France, John Paul Jones found his own vessel in such a deplorable condition that he saw it would take long weeks to put it once more in condition for service against the enemy.

During this time, he was thrown upon his own resources by the poverty of the Continental government and the commissioners' dishonoring of his draft.

In order to provide food for his crew and supplies for his ship, he was forced to sell his prize.

This act so displeased his government that he was asked to hand over the command of the *Ranger* to her first lieutenant. So when the ship finally sailed for home that fall, Jones was left in a foreign land with no prospect of a ship to rest his feet upon.

It was truly a cheerless outlook.

Furthermore, his treatment by Congress had much hurt the brave man's heart.

But he was not the kind either to grow bitter against his country or to grow hopeless.

John Paul Jones had two powerful friends in France.

These were the Duke de Chartres, eldest son of the Duke d'Orleans, and his wife, the Duchess de Chartres. Through them he was able to interest the King into providing him with a ship, the *L'Duras*, of forty guns. She was a very old craft.

After spending three months overhauling her, she was still not much better than a makeshift warship, for her batteries were mounted with the refuse guns of the French government—old, rusty, loose-jointed relics that had served their real usefulness and were liable to blow to pieces when fired with a heavy charge.

His crew, too, was a motley and ill-assorted bunch.

Of the three hundred and seventy-five men whom he was able to collect, only fifty were Americans; the rest were French, Portuguese, and British.

Fortunately, before the final date of sailing, the exchange of British and American prisoners of war gave him the opportunity to replace some of his alien seamen with one hundred and fourteen Americans.

As he stood and viewed the muscular bodies and keen eyes of these fellows, he breathed easier.

All arrangements were completed in August 1779.

By that time the *L'Duras*—whose name Jones had changed to *Bon Homme Richard,* out of compliment to Benjamin Franklin—had been converted into a fairly efficient looking man-of-war.

The *Bon Homme Richard, 1779* by artist F. Muller (1883–1966)

Besides the *Bon Homme Richard,* the squadron consisted of several smaller vessels, all commanded and crewed by Frenchmen.

There was the *Alliance,* Captain Landais, a frigate of thirty-six guns; the *Pallas,* a frigate of twenty-eight guns, and the *Vengeance,* a brig of twelve guns.

All excepting the *Alliance* belonged to the King, and French money paid the expenses of the expedition.

Yet the ships sailed and the men fought under the American flag, and the French officers were for the time being commissioned officers of the United States.

Commodore Jones himself was looked upon as an irresponsible adventurer by most of the French officers, who were very jealous of the praise he had won from their countrymen for taking the British frigate *Drake.*

No wonder, therefore, that there was a good deal of discontent.

The squadron finally set sail from at daybreak, on the 14th of the month. The projected cruise was to cover a circle around the British Islands and end at the Texel [an island off the coast of the Netherlands].

Through calms and gales and shifting winds, John Paul Jones worked his way to the west coast of Scotland, and then beat down the east coast as far as the Firth of Forth. Finally they reached Spurn Head.

In the meantime a number of seamen had deserted, most of the captains had shown a rebellious disposition, and Captain Landais had revolted to the extent of separating, with his ship, the *Alliance,* from the rest of the squadron.

In spite of these drawbacks a number of prizes had been taken and sent to friendly ports with prize crews.

On September 22nd news was brought to Jones by a passing friendly vessel that a large Baltic fleet of British vessels, laden with naval stores bound for England, had arrived in Bridlington Bay under convoy and was waiting for favorable winds to proceed on to the Downs.

Commodore Jones felt a great leap in his heart.

The moment he had passionately longed for had come at last. This was the finest kind of chance to hit England with a staggering blow.

Signaling his consorts to follow, he headed northward for Flamborough Head. The following morning found his squadron twelve miles at sea, just north of Bridlington Bay.

With his spyglass, he scanned the harbor. He saw moving ships, headed outward.

If his heart had leaped before it now fairly jumped out of his broad chest.

Still watching, he observed the Baltic fleet sail out of the bay, and hugging close to land, scamper for the shelter of Scarborough.

Instantly up to the masthead of the *Bon Homme Richard* went the signal for the chase. Then, noting their enemy after them, the merchant ships of Britain crowded sail, and their two escort ships moved out menacingly, like mother hens, to protect their retreat.

Commodore Jones was far more eager for a fight than they.

At once he accepted the challenge, changed his course slightly, and bore down upon the British warships.

The *Alliance*, which had for a long time been keeping at a distance, paid no attention to Jones's signal to form a battle line.

On the contrary, she stubbornly kept her place, and even had the audacity to signal the little *Vengeance*, "Lie to as you are; you are not big enough to bear a hand in this."

The *Vengeance* obeyed the order of the rebellious Captain Landais, thus leaving only the *Pallas* to follow after the *Bon Homme Richard*.

The British escort ships were the *Serapis*, a new forty-four-gun frigate, and the sloop-of-war *Countess of Scarborough*.

Captain Cottineau, of the *Pallas*, gave chase to the sloop, which was running out to leeward [direction in which the wind is blowing], and during the coming fight he was so fully occupied in capturing her that he could render no assistance to his commander-in-chief.

Commodore Jones was left single-handed to face the formidable Britisher *Serapis*—left to fight it out on an old rotten hulk that should have been made into firewood long before it ever came into his hands.

To make matters worse, his crew was foreign and unskilled and uncooperative, while Captain Pearson of the *Serapis* possessed men of the highest degree of training, perfectly organized.

That this fight resulted in a victory for Jones is one of the most remarkable tributes to the leadership and fighting qualities of the man that could ever be paid him.

Let us see how it all came about.

The *Bon Homme Richard*, on the left, fighting the *Serapis*, Sept. 22, 1779

It was seven o'clock in the evening before the two vessels came within striking distance. As the *Bon Homme Richard* came up, Captain Pearson sent a shot whistling toward her.

There was no answer.

After scanning her in the fast dimming light with his night glasses, he said to his first-lieutenant: "It is probably John Paul Jones. If so there is work ahead for us!"

Then he hailed again—with another whistling shot.

This time there was an answer.

The darkness about the stranger was split with fiery forks of flame; there was a terrible roar—and a broadside sent its iron pellets toward the *Serapis*, several of them striking their mark. Captain Pearson was now quite positive of the identity of the stranger's commander.

By this time the ships were within six hundred feet of one another. A light but steady wind was blowing from the southwest. The sea was smooth and comparatively peaceful.

Soon both ships were engaged in the hottest kind of firing.

Broadside sent its challenge, was accepted, and answered by broadside.

The surroundings were almost constantly bathed in the pale, livid light of flashing gunpowder, the contesting ships standing out like grim, black specters of the deep.

The night air was split and shattered by the crash of cannon, the roar of mortar, the rattle of musketry, the crash of mast, the splintering of hull, the wild and triumphant cries of men, and the hoarse shouts of alarm and death shrieks of other men.

In the midst of it all, there came a rending roar, more portentous, more dreadful of inflection than all other sounds, from the direction of the *Bon Homme Richard*.

The inevitable had transpired.

Riven with rust pits, weakened from long usage, two of the *Richard*'s old eighteen-pounders had burst.

They had formed a part of the steerage battery, just under the main deck, aft. The explosion killed or wounded most of the gun crews, and demoralized the crews of the four remaining eighteen-pounders, who refused to risk their lives any longer in firing them.

Thus, at the very beginning Jones found himself greatly handicapped anew.

A quiet word to his pilot, and the *Bon Homme Richard* now began to drift closer in upon her antagonist.

As the distance between them decreased, the deadliness of fire of both vessels increased.

Now pistols could be used; and the din of their rattle was added to the other fulmination of sounds. The rotten timbers of the American ship were a veritable pin cushion for the projectiles of the British gunners. They stuck in the old hull everywhere, and in places, great gaping holes appeared.

But the *Serapis* was far from escaping her share of the damage. Her hull, too, showed many marks of well-aimed shot; her decks were littered with dead and wounded.

The ships drifted slowly apart, still fighting madly. Lead rained down upon both crews as if the heavens themselves had turned storm wrath into bullets.

Nine of the *Richard*'s twelve-pounders had been abandoned; eighty men of the main battery had been struck down. The condition of her deck was terrible to behold. The wreckage of rigging and cabin was scattered everywhere; stark, red-stained bodies lay here and there; wounded, with agonized expressions and moaning lips, propped themselves against every conceivable support; the living and unhurt, with shirt-stripped bodies black from gunpowder, sprang wildly from point to point, loading and firing their pieces like madmen.

Matters below decks were even more alarming.

The hull had been pierced by several heavy cannon balls, and the water was pouring in very fast. Already four feet deep in the hold, it was steadily increasing.

Noting this situation, Commodore Jones hurried up to his first lieutenant, Richard Dale.

"Dick," he said, "his metal is too heavy for us at this business. He is hammering our old wood hull all to pieces. We must attempt to get hold of him again!"

It was really his last chance, his only hope.

By a skillful maneuver, he rounded the bow of the *Serapis* and suddenly came in upon her.

The enemy ship's jib-boom [an extension of the bowsprit, protruding from the front of the ship] ran over the *Bon Homme Richard*'s poop-deck [a partial deck that forms the roof of a cabin built in the ship's rear, or "aft"].

Before it could get away, two of the American sailors had seized it, and Jones himself jumped up and lashed it fast to his ship's mizzenmast [the third mast from the front, or bow].

This brought both ships together, side by side. Their cannons all but touched. The starboard anchor of the *Serapis* was hooked in the *Richard*'s mizzen chains. It was a secure lock.

The rest of that furious fight was a hand-to-hand struggle to the very death.

At this juncture, some of his officers came up to John Paul Jones, and attempted to persuade him to strike his colors [meaning, lowering the flag to signal surrender].

As John Paul Jones was about vehemently to reply, an under-officer, crazed with fear, shouted loudly to the British frigate:

"*Serapis* ahoy! Will your commander grant us quarter?"

[A call for quarter meant to cease firing and show mercy.]

The indomitable Scotchman's face turned red with anger at this flagrant breach of discipline and this show of weakness.

Only the hail of Captain Pearson saved the cowardly officer from being knocked to the deck.

"Does your commander himself call for quarter?" asked the British captain.

Glancing up to his masthead, Jones was surprised to realize for the first time that his ensign [an officer] had been shot away.

Stepping forward on the poop-deck, he called back: "Quarter nothing! We have just begun to fight!"

Captain Pearson was stunned.

Coming from any other man, he might have thought this defiant reply mere bravado. But he had heard enough about John Paul Jones to know he meant every word of it.

Commodore Jones now sprang on the quarter-deck, and became at once the life and center of the defense.

He rallied his men at the battery, shifted over one of the guns himself, and directed the fire. Then he dashed among the French marines, cheered and exhorted them in their own tongue with his great voice, and even took the muskets from some of them to set them an example in good firing.

The sole chance of victory plainly lay in clearing the enemy's decks.

To do this, everything depended upon the unswerving aim of the sailors on the tops and the marines on the decks. If the crew of the *Serapis* should succeed in casting off the lashings that held the two ships together, the fate of the *Bon Homme Richard* would be sealed.

Himself directing all his energies to the defense of the grapples for a time, until the men understood the work, Jones next turned his attention to clearing the enemy's decks.

In a short time, the foe could no longer stand at his wheel or handle his sail.

It was instant death to any British sailor. Their forecastle was soon abandoned by officers and men alike.

At this stage—about half-past nine—Commodore Jones could see the *Alliance* coming up. He was filled with joy. With the aid of his sulking consort, who had probably repented, he could soon gain the mastery.

But it was a false hope.

To his surprise and indignation, Captain Landais drew up and calmly discharged a full broadside into the stern of the *Bon Homme Richard*!

Believing his rebellious officer must have mistaken him in the darkness for the enemy, the Commodore now displayed three lanterns in a horizontal line—a signal of his identity.

The *Alliance* moved around a bit, there was another flash of flame along her hull, a roar, and once more came the thud of crunching wood in the *Richard*'s hull, and several of her crew fell in their tracks.

Another swing, and another broadside came hurtling into the sister ship.

More men fell, a shot pierced the old hull below the water-line, and another leak was added to her too-plentiful troubles.

John Paul Jones stood aghast.

He could hardly bring himself to credit the Frenchman's treachery.

He probably pinched himself to see if he were not asleep and dreaming.

Slowly the terrible truth dawned on him—the *Alliance* had really turned traitor!

Instead of a friend, she too was an enemy!

The commander of the *Bon Homme Richard* smiled. But it was a grim smile, a smile that boded ill for somebody in short order.

The more enemies, the better he could fight.

Now let them watch out!

By this time, the condition of the *Richard* was indeed desperate. Fire had broken out in the lower deck, and the flames, eating their way hungrily through the splinters that were everywhere to feed them, licked their way toward the powder-magazine. Five feet of water were in the hold; the ship was slowly sinking.

And still, Jones's resources were not exhausted.

Although the enemy's upper decks had been cleared, the lower tier, being covered, were untouched. The Commodore now suggested dropping hand grenades through the *Serapis*'s main hatch into this lower tier.

For this purpose, Midshipman Fanning, acting gunner, and two seamen were ordered aloft into the maintop. Armed with two buckets of grenades and a slow match, these four men went to work.

The hatch was only partly open, leaving a hole not more than two feet wide. But the main yard-arm [part of ship's rigging] of the *Bon Homme Richard* overhung it. Fanning, who was known as a good thrower, was to do the casting.

His first two grenades missed. But the third one went fairly into the opening.

A terrible explosion followed.

The hatch of the British frigate was blown high into the air. More than fifty men were killed or maimed, and the after part of the lower tier was silenced.

Just as the crew on the *Bon Homme Richard* were wildly cheering this exploit, the *Alliance* bore down for the second time upon her sister ship.

Coming up within musket shot, she again raked the shattered and sorely distressed vessel that was flying her own colors.

Fired at alike by friend and foe, it seemed the height of folly for Jones to persist, but he did. It was the fighting strain in his Scotch blood, inherited from his Highland mother's male ancestors, that made him fight on and on, knowing no defeat, till death might claim him.

Directing one of his gunners to aim for the mainmast of the *Serapis*, Jones soon had the satisfaction of seeing splinters fly and the mast waver. Another shot, and it sloughed, crashing into the sea, carrying with it rigging, and throwing up a geyser of spray far above the deck.

Already the Commodore stood by with a chosen party of boarders.

Armed with pistols and cutlasses, every man was ready for the last act.

Jones now gave the signal, and the men, shouting loudly, and under the command of Acting Lieutenant John Mayrant, plunged over the hammock netting and down into the fore-hold of the *Serapis*.

Meeting with little opposition, they were soon in full possession of the ship, and rushed toward the quarter-deck.

Seeing the hopelessness of resistance, Captain Pearson struck his colors [remember what that means?].

Not knowing this, his first lieutenant rushed up a moment later and asked, "Has the enemy struck, sir?"

"No, sir; *I* have struck," replied Pearson with tears in his eyes.

It was the first time the British flag on a man-of-war was ever struck to the Stars-and-Stripes!

Although the enemy had surrendered, there was still much work to do for the crew of the *Bon Homme Richard*.

Victorious as she was, she was rapidly sinking, her old wounds having grown larger.

Seven feet of water filled her hold. The pumps, still manned by the British prisoners, were beginning to choke.

Of her former crew only a hundred men were left unwounded, nearly every gun had been dismounted, her starboard [or right] side was completely destroyed, and the flames were breaking out afresh, threatening to consume her or cause her magazine to blow her into a thousand bits.

One alleviating circumstance was that the profligate *Alliance*, having observed that the crew she so cravenly hated had won the fight, had slunk away in the darkness and no longer contributed to the distress of the *Richard*.

As may be surmised, the crew of the latter ship had very little sleep the remainder of that night. When another day dawned, all of the wounded were removed to the *Serapis*. Before any of the stores could be saved, the ship began to list badly, and it was deemed dangerous to remain any longer on her.

All that night, however, the old shot-torn craft kept proudly afloat. Too much glory surrounded her for her to perish when eyes could not see.

The first golden rays of next morning's sun still found her afloat. They caressed her pitiful ragged hulk into a brightness of unusual nobility and charm, and bathed the faces of the dead, lying in long tiers on her deck, in a holy radiance.

Suddenly she seemed to shudder. Then her head was seen to slowly bend.

Lower and lower the ship sank, till the waves were lapping close to her jib-boom.

There was a momentary pause, as if she were halting to take a last look at the scene of her sacrifice, when, with a rare dignity, she went down.

As she plunged into the great depths awaiting her, her taffrail [the handrail around the deck] momentarily rose in the air. And the very last vestige mortal eye ever saw of the *Bon Homme Richard* was her waving, tattered flag, still unstruck!

CHAPTER IV

LORD HORATIO NELSON

His few surviving comrades saw
His smile, when rang their proud hurrah,
And the red field was won;
They saw in death his eyelids close, Calmly, as to a night's repose,
Like flowers at set of sun.

—*Halleck.*

A Great Sea King

THE DEEDS OF Lord Nelson, Britain's greatest sea hero, form a dramatic chapter of life that should be like a torch to fire the admiration of every boy in the land, no matter from what country he hails. Nelson's exploits should prove especially stimulating reading to the lad who is physically frail and has the erroneous idea that only robust boys and men ever excel in feats of red-blooded daring.

To have looked at Horatio Nelson's thin, pale face and slender body—particularly to have known of his frequent illnesses—you would never have connected him with great accomplishments requiring the snappy action of a healthy, vigorous mind, and the equally snappy action of a fearless, obstinate fighter.

The development of this boy into England's foremost sea-fighter astonished his youthful playmates and acquaintances quite as much as you, or one of your young friends of a "weak constitution" can sometime surprise your boyhood comrades.

Nelson's life was a perfect drama in its happy rise, its glorious course, its heart-stirring ending.

The spot-light struck his figure when he was thirty-nine; at forty-seven his stage was steeped in darkness.

But in those eventful eight years, he had won his trio of marvelous triumphs—Aboukir, Copenhagen, Trafalgar—made Great Britain "mistress of the seas," and left a memory of such splendor behind him that his name will ever be dear to his countrymen and known to all future generations of reading peoples.

This great man was born on the 29th of September, 1758, at Norfolk, on the eastern coast of England. His schooling was scant. Although of a weak and sickly constitution, his ambition from boyhood was to be a seaman, and when in 1770 he heard that his uncle, Maurice Suckling, had been appointed to the command of the *Raisonable,* he begged to join him.

On this ship Nelson sailed to the Falkland Islands, being then just twelve years old. A little later he made a year's voyage on a West Indian merchantman, on which he learned the essentials of his profession.

In 1773, an expedition was fitted out by the Royal Geographical Society to sail to the North Pole. Young Nelson, then only fourteen, was wild to go. As no boys were allowed by the Admiralty he almost failed in this endeavor, but finally prevailed upon Captain Lutwidge, of the *Carcas*, to let him be a member as coxswain [the swain (boy servant) in charge of the small cockboat used to row the captain to and from the ship].

When Nelson returned from the ice-bound seas the next year, he was a full-fledged able seaman.

Then came a three-year voyage to the East Indies under Admiral Hughes.

This cruise, along fever-stricken, marshy shores, undermined his none-too-rugged health. A long sickness ensued which came very near to resulting in his end. For at least one moment in his life, he felt an over powering despondency. He saw before him the towering obstacles he must meet in his chosen profession, and the tiny, puny, physical strength he had with which to meet them.

His ambition consumed him, but his mind could see no way to satisfy it. This is how he afterward refers to it: "After a long and gloomy reverie, in which I almost wished myself overboard, a sudden flow of patriotism was kindled within me, and I saw my King and countrymen very much smaller and weaker than I, needing my aid. My mind exulted in the idea. 'Well, then,' I exclaimed, 'I will be a hero, and, confiding in Providence, I will brave every danger!' "

In those early years, and throughout his entire lifetime, duty and fame were the two thoughts that appealed to him most strongly, the two thoughts that constantly spurred on his endeavors. Duty kindled in him the most fearless and unswerving devotion to his fellowmen; renown lured him ever onward with its radiant, glittering honors.

Nelson's promotion was swift. In July, 1778, he won a first-lieutenancy on the *Bristol*, the flagship of Sir Peter Parker. Six months later he was appointed commander of the *Badger*, a brig. Then came a change to post captain of the frigate *Hinchingbrook*, with which he proceeded against American privateers in the West Indies.

Finally, on February 1st, 1793, France declared war against Great Britain and Holland. Two days before, anticipating trouble, his government had appointed Nelson to the *Agamemnon*, a sixty-four-gun ship.

H.M.S. *Agamemnon*

During the ensuing four years, Nelson took part in many minor engagements, all helpful to his cause. Among these were the investments of Marseilles and Toulon and the blockading of Bastia. In the latter action, while serving the battery, an enemy shot drove a quantity of sand into his right eye, which eventually destroyed its sight.

The chief event marking the opening of 1797 was the first important naval operation connected with his name. This was the battle of Cape St. Vincent, fought against the Spaniards, the allies of France.

At dawn, on the 4th of February, the British fleet of fifteen ships, under Sir John Jarvis, lay twenty-five miles west of Cape St. Vincent on the southern coast of Portugal.

A heavy fog darkened the morning. Soon after eight o'clock, the fog lifted, and there, coming toward them, the lookouts on the British vessels saw the grand fleet of Spain. The sight was enough to awaken apprehensions, for there were twenty-seven huge ships.

Among these were the great *Santissima Trinidad,* carrying one hundred and twenty-six guns, and the *San Josef* of one hundred and twelve guns. One was a four decker—the largest type of ship afloat—and seven were three deckers. The Spanish ships were in two sections, arranged in a most awkward line of battle, seemingly unable to unite.

Meanwhile the British admiral made the signal to prepare. At eleven o'clock, the line was formed, and at eleven thirty, action began.

Nelson's ship was thirteenth in line. The plan of Sir John was to run between the two divisions of the enemy's fleet, thus preventing their junction, and then assail them separately.

When he had succeeded in dividing them in this manner, he made the signal to "tack in succession" [each ship tacking, or turning its bow toward the wind, allowing more progress in a desired direction].

At this moment Nelson's chance came.

The Moonlight Battle off Cape St Vincent, 16 January 1780 by artist Richard Paton

Seeing that it was the intention of the Spanish admiral to run behind the British column and unite his lines, Nelson, without orders, passed from the rear, between the *Diadem* and *Excellent*.

Reaching the Spanish column ahead of his compatriots, he engaged single-handed the huge and formidable *Santissima Trinidad*.

Shortly, the *Culloden*, leader of the British van, came to his support, and for an hour the two plucky British ships had their hands full; for by this time nine other Spanish vessels had come to the aid of their great consort.

Then the pair of British boats were joined by the *Blenheim* and the *Excellent,* and the enemy craft *San Ysidro* and *Salvador del Mundo* were compelled to strike their colors.

Nelson's own ship, the *Captain,* had suffered considerable damage. There was not a sail, shroud, or halyard [rope used to hoist a sail] left intact. Her wheel had been shot away. She was quite incapable of making chase, or fleeing under chase.

But she had neither to do just now. She had been worked up alongside the *San Nicolas,* and the next instant Nelson ordered a party to board her.

Captain Berry was the first to leap to the Spaniard's deck. He was quickly followed by Nelson himself. A few volleys of musket and pistol shots were exchanged, then the enemy backed away. Berry gained possession of the poopdeck and hauled down the Spanish colors. At the same time Nelson, on the forecastle, received the swords of the Spanish officers.

Just at this time the *San Josef* came to the aid of her distressed sister ship, firing on the British aboard the *San Nicolas.*

Nelson promptly called for reinforcements from the *Captain.* Without waiting for these, the intrepid Englishman ran across to the farther side of the enemy's deck and sprang to the rail of a large three deck Spanish ship which lay close to that quarter.

As he gained the deck of the second enemy craft, he met a Spanish officer. Apparently believing that the British had boarded him also and that resistance was hopeless, this officer at once gave his sword to the dauntless Nelson, who, quite dumbfounded, received it and listened to the Spaniard's words, "We surrender."

A little later the *Victory,* Admiral Jervis's flagship, passed the disabled *Captain* and its two magnificent prizes, and saluted with three cheers, as did every ship in the fleet.

On the 14th of July, 1797, Nelson, now a rear-admiral, received orders to sail for Santa Cruz. Information had come that a Spanish ship, the *El Principe Asturias*, was in this harbor, and would sail shortly to Cadiz with a rich treasure belonging to the crown of Spain which she had brought on from Manila.

Nelson's squadron consisted of four warships and three frigates, carrying a landing party of a thousand men under Captain Troubridge, and one cutter [a small warship that can use sails or oars].

He had planned a sudden and vigorous attack. Upon arrival, the boats, with troops and scaling-ladders, were ordered to land in the night. But a heavy gale of wind and a strong current prevented it.

The next day they hoped to storm the heights behind the forts, while the ships discharged their batteries in front. But when morning came calms [no wind for the sails] and contrary currents kept the large ships out of range. This condition prevailed for several days.

By the 24th of July the Spaniards, who could readily see the British ships, had had four days in which to strengthen their works and increase the number of their troops.

Nelson straightway made up his mind to do something at once. That day he wrote to Jervis: "To-morrow my head will probably be crowned with either laurel or cypress." [Crown of life or crown of death.]

There was, indeed, small hope for success.

Nelson in the Captain engaging Santissima Trinidad and San Josef by artist Alma Claude Burlton Cull.

That night, about eleven o'clock, the boats advanced quietly toward the town with seven hundred and fifty men. Nelson's barge led the way. Hardly had they proceeded to within half-gunshot than they were discovered. A sharp fire of cannon and musketry opened upon them from one end of the town to the other.

The sky was dark, and the surf high. A number of the boats missed the landing; they were full of water in a minute, and crashed upon the rocks. The ladders were all lost in the surf; the ammunition was wet and useless.

But the other barges reached the position successfully. Led by Nelson, the British stormed and drove back the five hundred Spaniards defending the city and prepared for further advance.

However, such heavy fire came from the citadel and town that the British began to fall on all sides. They were forced to retreat in the face of the terrible shower of bullets.

Already a grapeshot had struck Nelson in the right elbow. His stepson, Josiah Nisbet, had placed him in the bottom of his boat, bandaged the wound tightly, and rowed back to the ships.

As they neared the ships, a cannonball struck the cutter *Fox* beneath her waterline, and she went down. Although suffering intense pain from his wound, Nelson insisted on stopping to save the men struggling in the water.

When they finally reached the *Theseus*, a rope was thrown over. With a rare display of fortitude, Nelson climbed the rope to the ship's deck, and immediately called the surgeon to bring his instruments, for he knew his arm must be amputated.

His feelings when he returned to his home shores after this mishap, with the defeat at Santa Cruz also on his mind, must have been bitter.

To add to his troubles, the amputation had been crudely done, and for long weeks he suffered tortures from the imperfect surgery.

Yet it was after this when he went forth again—a one-eyed and one-armed fighter—that the glorious luster of his renown attained the pinnacle of its height to burn undimmed to the end.

The Fight of the Nile

FRANCE HAD MADE peace with most of the continental powers and had extended her influence over all adjoining countries.

In the spring of 1798 she was concentrating her forces against Great Britain with the avowed purpose of destroying the British monarchy.

Active preparations were being urged forward in the seaport towns of the Mediterranean, both at the southern ports of France and at the friendly ports of Italy. Ship-of-the-line, transports, and troops were assembling in large numbers. An extensive naval expedition was evidently being planned.

This operation, which afterward proved to be the famous expedition of General Napoleon Bonaparte to Egypt, was kept such a profound secret that its destination was not even surmised by the British.

But the British government was sufficiently alarmed to cause it to decide to abandon its purely defensive tactics and to assume an offensive attitude.

At this stage Nelson, in command of the *Vanguard*, a seventy-four-gun ship, was off Cadiz with Lord St. Vincent. The latter now gave him two ships of the line, the *Orion* and the *Alexander*, and four frigates, and sent him to watch Toulon and follow the movements of the French fleet.

From off Cape Sicie, Nelson soon after reported that nineteen sails lay in the harbor of Toulon, that transports with troops frequently arrived from Marseilles, that twelve thousand men were already aboard the warships, and that hearsay had it the fleet would sail for an unknown port in a few days.

Unhappily, the French fleet succeeded in slipping out of the harbor without Nelson detecting the fact until too late. This was owing to a strong gale dismantling the *Vanguard* and rendering it necessary to tow her to a port on the coast of Sardinia, Italy ,for repairs.

The same storm drove the other vessels far apart, and precious days were lost in reassembling them.

Thus it was not until the 7th of June that Nelson, bolstered by the addition of ten other ships from St. Vincent, took up his famous pursuit of Bonaparte.

He had been ordered to attack him wherever found.

The French had a long start, and all that Nelson knew of them was that they had sailed southward between Italy and Corsica, and had been seen by a passing vessel off the north end of Sicily, steering to the eastward.

With this clue, the British squadron made off under a full spread of canvas. As they went along, from time to time they picked up additional fragments of news regarding the chase, which helped them shape their course.

It was a long and tedious search. There were times when they expected to come within sight of the enemy at any moment; other times when they almost despaired of ever seeing him. The want of fast frigates hampered Nelson a good deal, his heavy warships not having the requisite speed for a stern chase of this type.

On the 15th of June, he learned from a Tunisian cruiser that the French had been seen off Trapani in Sicily; on the 26th news came off Cape Passaro that the enemy had arrived and possessed themselves of Malta, and then gone eastward.

From this Nelson was now pretty well convinced that Bonaparte intended to take some port in Egypt, establish himself at the head of the Red Sea, and carry a formidable army into Hindustan.

If this surmise were correct, British interests in India were in great danger.

But if to Egypt, where in Egypt had they gone?

After a consultation with his most trusted captains, Nelson decided to head for Alexandria.

Cruel suspense marked the next few days of the chase. Every moment, night and day, the lookouts in the foretops were at their posts, hoping to be able to report the first sight of the French. But disappointment still prevailed.

Alexandria was reported in view on the 10th. Not a French sail was to be seen, nor could any information be gathered of the whereabouts of the other fleet.

Nelson was much downcast, even mortified. His judgment, for once, had played him a trick.

He did not know that he had actually outstripped his adversary; that his unerring reasoning power had led him to Alexandria before Bonaparte had arrived. The truth was, the French fleet had veered to the south shore of Candia [now part of Crete, Greece], and strange as it may be, under cover of the night and a dense fog, the two rival forces had at one time been within gun range of one another without knowing it!

Battle of the Nile, August 1st 1798 at 10 pm by Thomas Luny

Not aware of this, Nelson now stretched his ships over the coast of Asia, then steered along the northern shore of the Mediterranean, passed Candia once more, and returned to Syracuse sorely disappointed and humiliated, after a cruise of twenty-seven days.

On the 23rd of July, not to be thwarted, Nelson again set out in quest of the elusive French.

Once more he sailed eastward and southward, still convinced that Egypt was the goal of the enemy but puzzled at his inexplicable disappearance.

Six days later he sighted the Pharos [lighthouses] of Alexandria—and there, in Aboukir Bay, fifteen miles from the port, rode the French fleet in solid battle array!

Nelson was not found unprepared.

Throughout that long chase, it had been his custom to hold almost daily conferences with his captains, during which plans for attack or defense under every conceivable condition had been worked out. Therefore, it took but the briefest word now for the commander-in-chief to acquaint his officers with their part in the problem that confronted them.

It is from the extended accounts of Captain Berry of the *Vanguard* and Captain Miller of the *Theseus* that we know the details of this wonderful battle fought in the inky darkness of the night, in waters and among islands and headlands entirely unknown to every officer in the British fleet.

When the French first saw him, the enemy was nine or ten miles to the southward, with Aboukir promontory and island and a network of dangerous shoals and reefs lying between.

The keen eye of Nelson saw the weak point in the enemy's position; he noted that where there was room for a French ship to swing, there was also room for a British ship to anchor.

By taking up positions inside and outside the French line, he could concentrate his fire on the enemy's vanguard and center, while the wind would prevent the French rear from coming to the assistance of their consorts.

Battle of the Nile—Night of 1st August 1798, engraved by Thomas Hellyer.

Circling around Aboukir point, and giving the shoals a safe berth, the British ships advanced upon the enemy in a single column.

First came the *Goliath*, followed in order by the *Zealous*, the *Orion*, the *Audacious*, and the *Theseus*.

The *Vanguard*, Nelson's flagship, was sixth in line. A few minutes before sunset, the *Goliath* and the *Zealous* opened up with their heaviest guns. Ten minutes later Nelson gave the signal to engage the enemy at short range.

Advancing with silent guns, the British ships could be seen to have many sailors aloft who were furling sails and hauling in braces, preparatory to casting anchor.

As they swung in to take their positions, the concentrated fire of the French broadsides was projected into their bows.

Then with swift movement and masterly daring, the *Goliath* and the *Zealous* worked around the enemy's flank. Although deluged by the fierce, raking fire, they passed inside the ship's line.

A moment later the inshore side was taken by the *Orion*, the *Theseus*, and the *Audacious*, while the *Vanguard* and four other vessels pitched into the French front and center on the outside.

Considering that they had been thus cleverly subjected to two fires, the French fought with admirable courage. The *Theseus* and the *Guerrier* engaged in almost a grapple, not six feet apart. The guns of the former, loaded with two and three round-shot, belched their flame and hot lead directly into her adversary. In the twinkle of an eye the mizzenmasts [the third mast from the bow] of the British ship fell by the board, her foremast having gone before.

A little way off, twenty minutes later, the *Conquérant* and the *Spartiate*, both French, were also dismasted.

In the very face of success, Nelson was struck down. A bullet had hit him in the forehead, blinding him completely for the moment. Quickly, he was carried below in the arms of Captain Berry, and a surgeon summoned.

After his wound had been dressed, and while he was still suffering intensely, he groped for a pen and paper and scrawled the first words of his dispatch to St. Vincent announcing the victory.

The Victory at Trafalgar

FOLLOWING THESE RECORDED events, Nelson had gained a brilliant victory at Copenhagen, been raised to the rank of viscount, and in October 1801, took his seat in the House of Lords [the upper house of British Parliament].

About the same time cessation of hostilities with the French republic was announced, and the following March a treaty of peace was signed.

Then, just a little over a year later—in May 1803—war with France broke out afresh.

Nelson was appointed commander-in-chief of the King's ships in the Mediterranean.

On the 18th he hoisted his flag on board the *Victory*, the ship whose name was to become almost as immortal as his own as the one in which he fought his last battle and won the crowning triumph of his life.

Meanwhile at Toulon and Brest, the French navy was being daily increased and put into effective condition. New ships were equipped, troops gathered for embarkation; every preparation was being made for a fresh naval expedition of large proportions—none less than the invasion of England.

For this purpose, Napoleon had been drilling an army of one hundred and fifty thousand men, and collecting a fleet of thirty-five ships. The squadrons were to assemble at the West Indies and, from there, sail for the Straits of Dover, England.

Cognizant of the assembling of the squadron at Toulon, Lord Nelson joined Sir Richard Bickerton and, on the 8th of July, began the blockade of that port.

It was the most tedious kind of work, this effort to keep the French vessels within from getting out, and keeping those from without getting in.

But through fair weather and foul, calms, winds, squalls, and gales, the blockading was faithfully carried on, although only four of his ships were strictly seaworthy craft.

Nelson's wisdom showed him the importance of preventing the junction of the French squadrons, and he made up his mind to maintain his vigil if it should take the balance of his life. If the French ships did attempt to break through him there at Toulon he would give them battle.

This is exactly what the French presently undertook to do.

After being bottled up by the British for fully a year and a half, their fleet, under Admiral Villeneuve, slipped quietly out of Toulon harbor on the night of the 17th of January, bound for Sardinia.

Nelson's two lookout frigates signaled him of the enemy's act two days later.

Within three hours Nelson was at sea in hot pursuit.

He scoured the Mediterranean in the worst weather he had ever seen; gale after gale tossed his ships over the angry waters. It seemed that the storms never would abate.

Now it happened that the same ill weather that tormented Nelson and added to the perplexities of the pursuit created havoc with many vessels of the pursued. So battered did they become that Villeneuve was driven to return to the port he had just quitted in order to repair them.

When the craft had all been refitted, Villeneuve again set sail.

Meanwhile Nelson, unaware of the enemy's procedure, had been battling and drifting with the storms. He had been to Naples, Sicily, and Sardinia; to Alexandria, Candia, and Malta; he had "covered the channel from Barbary to Toro with frigates and the fleet."

Yet—of course—he saw nothing of the French fleet.

It was deeply mortifying to Nelson to be thus outwitted; he ached immeasurably under the humiliation. He made up his mind, finally, to go out of the Mediterranean, working westward.

Storms and unfavorable winds continued to dog him. At this time, he wrote in his diary: "My good fortune seems flown away; I cannot get a fair wind. Dead foul!—dead foul!"

Through the Straits, past Cape St. Vincent, then straight across the Atlantic to the West Indies, the British fleet struggled on. With Nelson at the head, there was no such thing as giving up. Then came the Barbados and fresh rumors that the French fleet had been seen in Caribbean waters.

On he plowed to Tobago, to Trinidad, to Montserrat, to Antigua.

Had the enemy been at the latter place? Yes; but he had gone.

Once more the French fleet had slipped through Nelson's fingers. He grimaced, thought of his long and similar chase after Bonaparte, of its successful outcome, and—kept doggedly on.

Following rumors as to the direction taken by the French, he now turned about and headed across the Atlantic toward Cadiz.

On the 18th of July, three months after he had left the Strait of Gibraltar and had covered nearly seven thousand miles of sea, he sighted Cape Spartel [off Morocco].

But there was no French fleet.

Two days later, he reached Cadiz, and went on shore for the first time in two years.

A few days following he heard that the enemy had been seen steering northward.

In that direction he went.

But it was only to find nothing.

Receiving orders to return to England, he sailed for Spithead.

His stay was short. Two weeks after his arrival he sailed from Portsmouth to Cadiz, where he joined Vice-Admiral Collingwood. The force under his command then counted twenty-seven warships. In the harbor were the combined fleets of France and Spain under the command of Villeneuve and Gravina.

Villeneuve at last!

With the enemy were thirty-six warships and a number of frigates, apparently all ready for sea.

On the morning of the 19th of October, Nelson was cruising off Cape Trafalgar, keenly alert to any movement of the enemy. It was a clear day, with an easterly wind.

All at once the signal flew up on the British lookout ships that the French were coming out of port.

Seeing this, Nelson ran up the signal for a "general chase."

The effort was to prevent the enemy from entering the Mediterranean.

Toward the Strait of Gibraltar went the British ships, tearing along under full sail.

Two days later—at dawn of Monday, October 21st—the whole French and Spanish fleets had put to sea and were formed in a curved line of battle, stretching five miles from tip to tip, off the southern coast of Spain.

On one side lay Cadiz; on the other, Cape Trafalgar; in the far distance was the Strait of Gibraltar, through which the British were determined that the enemy should not escape.

Evidently he had determined the only alternative left for him now was to fight it out.

The Battle of Trafalgar

Artist J.M.W. Turner's *The Battle of Trafalgar* shows the *Victory*'s mizzenmast falling, the *Achille* on fire in the background, and the *Redoutable* sinking in the foreground.

TOWERING HIGH AMONG the craft of the Allies was the great Spanish ship, *Santissima Trinidad*, the largest vessel afloat and carrying one hundred and twenty-six heavy guns—a ship we have met before.

Directly next to her rose the masts of the *Bucentaure*, the famous flagship of the commander-in-chief, Admiral Villeneuve. On either side of these, Spanish and French leader-crafts stretched the extensive wing of their consort, bristling with guns, ready for the conflict.

For Nelson, the moment sought for two years and four months had at least arrived.

Appearing on deck of the *Victory*, dressed in his admiral's coat, and covered with a blaze of decorations which were the pride of his heart, he made in quick succession the signals:

"Form the order for sailing."

"Prepare for battle."

"Bear up."

Beating round, the British now came forward in two columns to the attack.

Admiral Collingswood in his flagship, the *Royal Sovereign*, headed the column to the south, while the *Victory* led the northern ships.

Toward eleven o'clock, Nelson went below, and on his knees wrote the words of his noble prayer.

"May the great God whom I worship grant to my country, and for the benefit of Europe in general, a great and glorious victory. Amen."

He followed this almost immediately afterward with the memorable message that was the last to his fleet.

"England expects every man will do his duty."

Shouts and cheers along the whole line greeted the inspiring words.

Directing her guns full upon the Spanish line, the *Royal Sovereign* began firing as soon as she was within range.

At the same time the *Victory* swept down upon the *Bucentaure*, on the French wing. As Nelson drew nearer the whole artillery of the French vanguard was directed upon him in a staggering violence. Then he opened up on the *Bucentaure* with such a terrific retaliatory fire that twenty of her guns were speedily dismounted and four hundred of her men killed or wounded.

Leaving the French flagship to his consorts, Nelson now turned his attention to the *Redoubtable*, which had already been attacking him. Running up close to the second French ship, he ordered the guns on that side to be fired. As they spat out their flame and shot, the *Redoubtable* shivered, for the range was so close that the guns of both vessels were almost touching.

Nelson and Captain Hardy paced the quarter-deck of the *Victory*.

With satisfaction, both noted the effectiveness of the British fire.

Almost above them, fifty feet up, the mizzentop of the rival craft swarmed with the best of France's sharpshooters. They had been told by their commander to "bring down Nelson—the officer with the medals" if ever the chance presented itself.

Now that opportunity was theirs.

There he stood below them in clear, unobstructed view, hidden only at moments by the shifting clouds of battle smoke.

A number of French muskets were quickly raised in the enemy mizzentop. The muzzles were trained downward at the calm and unconcerned one-armed admiral.

Just then a number of his men noted his peril—saw the Frenchmen aloft—saw their deadly intent. A warning cry was raised, and a number of British bullets sped upward and found marks.

But it was too late.

As Nelson wheeled about, several of the French sharpshooters' guns cracked, and Nelson, shot through the back, fell to the deck.

Captain Hardy quickly stooped over him. "They have done for me at last," gasped the great admiral. "My backbone is shot through!"

He was carried below to the cockpit, among the wounded and the dying, where everything was done to relieve his suffering.

For three hours he lay, listening intently to the incessant strife going on overhead in an effort to detect which way the battle seemed to be going, while his own life moments ticked rapidly away to the zenith of the dial. Of attendants and wounded alike, as they appeared, he asked weakly, "Are we still winning?"

Before he lapsed into unconsciousness he knew that the *Bucentaure* had been taken; that while his flag was still flying, seventeen of the French and Spanish ships had been captured, and one of the most glorious of British sea victories had been virtually won.

Even as he was closing his eyes for the last time, the roaring guns fell into peaceful silence, and everything was hushed and quiet when he whispered his final words.

"Thank God, I have done my duty!"

CHAPTER V

THE BURNING OF THE *PHILADELPHIA*

Burning of the Frigate Philadelphia in the Harbor of Tripoli by Edward Moran

'Twere better thus to sink where it stood
The ship he loved than that it should
Be used by foe against the good
Of his countrymen.
Swift stroke of oar—then
Flames on the ship shot to'rd the sky,
And when the pirates saw the why
They raged with mad, defeated cry.

—*Curtis.*

IT IS VERY difficult for us today to conceive that there ever was a time when the United States paid money to another country just to be left alone.

It is even more difficult to imagine the United States paying such a tribute to *pirates*.

And yet this is exactly what America did once upon a time.

Moreover, those same pirates—the sea robbers of the Barbary coast states of Morocco, Tunis, Tripoli, and Algiers, lying along the northern coast of Africa—succeeded by their indiscriminate and destructive raids on commerce in forcing most of the civilized powers of Europe to bend to their demands for a tribute in order to be left alone.

Even then peace did not come.

These disreputable corsairs of the seas—like the bully who takes the small boy's marbles in agreement to be friendly and then turns around with fists to demand more treasure—boldly continued to prey on Mediterranean merchantmen, to slyly deny their misdeeds, and to offer generous new assurances that all would be well if just a little more tribute were offered.

The only excuse that America can offer for ever entering into a shameful bargain of this kind and maintaining it as long as she did is that "everybody else was doing it."

For this purpose Congress appropriated money, and treaties were made by the President and ratified by the Senate.

On one occasion, at least, Congress actually revoked the order for new ships for the navy, and took the money to again buy off those pests of the seas, the Barbary buccaneers. The fund for this disgraceful purpose was known as the "Mediterranean Fund," and was entrusted to the Secretary of State to be disbursed by him at his own discretion.

Thus matters went on for some years.

But the time was soon to come when a summary stop was to be put upon the blackmailing—at least for this country.

Around 1800, it occurred to the American government that perhaps there was a more efficient and more manly way of dealing with the Barbary powers than feebly handing thousands of dollars over to them every time they took a notion to demand it.

This change of attitude was largely occasioned by a number of complaints which came in about this time from the blackmailers themselves. I think you can guess pretty well as to the nature of these grievances.

The Pasha of Tripoli whined that we had given more money to some of the Algerian ministers than to him; the Algiers government grumbled that we should not pay Tripoli so much; Morocco and Tunis pleaded for better treatment.

With nations, as with individuals, when the payment of blackmail is once begun there is no end to it. In the present instance, the situation was made even more acute by the declaration of the Pasha of Tripoli that he had ordered his subjects to cut down the flagstaff in front of the residence of the American consulate [diplomatic headquarters in the country], and was about to declare war against the United States.

This meant a renewed onslaught against American merchantmen by his pirates unless we immediately stepped forth with sufficient gifts to buy the old rascal off.

In answer, the United States government did a most fitting thing.

In 1801, Congress sent Commodore Dale with a small squadron of warships to teach the pirates a much-needed lesson.

No sooner had the expedition reached the Mediterranean than the cowardly Morocco and Tunis stopped their grumbling and came to terms with the United States.

This left Tripoli alone to be dealt with.

As Commodore Dale had sailed before the declaration of war with Tripoli was known, he was hampered by his orders, which permitted him to protect our commerce only, and which forbade hostilities.

Nevertheless he was destined to get into a fight, for hardly had he appeared off Tripoli than he was fired upon by the Tripolitan ship *Tripoli*, of fourteen guns.

No self-respecting, peacefully inclined American would ever stand that, and Dale's own guns were quickly barking away in defense.

Tripoli pirates attack US Lieutenant Decatur. His one hand is uplifted to stop a fatal knife thrust, while with the other presents his pistol. Near him lies his broken sword, and at his feet, a sailor used his body to shield Decatur. In the background, is Tripoli and American gun boats. Engraving by Whitney & Jocelyn.

The battle engagement lasted a matter of three hours.

Then the *Tripoli*, having suffered sufficiently for her temerity, struck her colors. Her mizzenmast was gone, she was riddled with the well-aimed shot from the Americans, and fifty of her crew lay wounded or dead.

Scarcely any Americans on the *Enterprise*—the twelve-gun ship commanded by Lieutenant Sterret which had fought this battle—had been wounded.

It was a signal victory.

As he had no orders to make captures, Sterret threw all the guns and ammunition of the *Tripoli* overboard, cut away her remaining masts, and left her with only one spar and a single jury sail to drift back to Tripoli—a hint to the Pasha of the new American policy.

In 1803 the command of the United States fleet in the Mediterranean was taken over by Commodore Preble, who had just succeeded in forcing satisfaction from Morocco for an attack made upon his country's merchantmen by a vessel from Tangier.

He was preparing to enforce a blockade of Tripoli when news reached him that the frigate *Philadelphia*, of forty-four guns, commanded by Captain Bainbridge, and one of the best ships in the United States navy, had gone upon a reef in the harbor of Tripoli.

It was said that the mishap occurred while she was pursuing an enemy, and that while she lay helpless, she had been surrounded and captured by Tripolitan gunboats.

This was a very serious blow. It not only weakened the American force in those waters, but it was also a great help to the enemy, for it promised to add another vessel to their own fleet.

In fact, the enemy got the *Philadelphia* off the rocks, towed her into the harbor, and anchored her close under the guns of their forts. They also replaced her batteries [artillery] and made every effort to prepare her for service against her own ships.

At this critical time Stephen Decatur came upon the scene to save the day for his countrymen.

Decatur, a young lieutenant of the *Enterprise*, appeared before Commodore Preble and offered to go into the harbor and destroy the *Philadelphia*.

Preble was willing, after talking at length with the intrepid young man and hearing his plans in detail.

However, carrying out the program was delayed, as the squadron was driven off the coast by severe gales. However, in January 1804, the favorable moment having apparently arrived, the Commodore assigned to Decatur a small vessel known as a "ketch" that had recently been captured from the enemy.

Decatur took seventy men from the *Enterprise* and put them on the *Intrepid*, as he now chose to rechristen the prize. Then, accompanied by Lieutenant Stewart in the *Siren*, who was to support him, he set sail for Tripoli.

The little ketch was very cramped quarters for so many men, but they were used to discomforts in those days and finally reached their destination without mishap.

For almost a week, they were unable to approach the harbor owing to severe gales that threatened the loss of their vessels. They ran into a hidden bay for protection.

But the weather moderated, and on February 16th, they stole out in the darkness of the night and made their way into the harbor.

All this time the *Philadelphia*, with forty guns mounted, double-shotted and ready for firing, and manned by a full complement of the pirates, was moored within half-gunshot of the Pasha's castle, the mole and crown batteries, and within easy range of ten other batteries which mounted a total of over a hundred guns.

Some Tripolitan 133 cruisers, two galleys, and nineteen gunboats also lay between the *Philadelphia* and the shore.

Leaving the *Siren* behind, into the midst of this powerful armament of the enemy Decatur had to go with his little vessel and only four guns.

But he was undismayed.

By seven o'clock he was stealthily making his way between the reefs and shoals which formed the mouth of the harbor. With the aid of a light wind blowing inshore, the ketch bowled silently along. Her master steered straight toward the *Philadelphia*, whose position he well knew and whose lights he could soon make out, like burning embers against the dark heart of night.

By nine thirty he was within two hundred yards of the former American frigate.

As they approached, Decatur stood at the helm with the pilot, only two or three men showing on deck. The rest of the crew of seventy-five were lying hidden under the bulwarks. In this manner the *Intrepid* drifted to within twenty-five yards of the *Philadelphia* before she was hailed by the lookouts.

The cry was in a foreign tongue whose meaning the young American could only guess at.

He answered back in English, hoping that it would be understood: "This is the *Intrepid*."

"You Angleesh?" came the broken inquiry through the darkness.

"We are alone," said Decatur, as though he had either not heard or had misunderstood the question. "We are in need of aid."

"What wrong?" was the next call. Evidently, the pirate's suspicions, if he had any, were allayed, as Decatur intended they should be.

"Our anchors have been lost in the gale," was the truthful answer. (The *Intrepid* and *Siren* had each lost an anchor.) "Will you let us run a warp [attach a line] to your ship and ride by her till morning?"

The truth of the matter is, while this little play had been going on, some of Decatur's men had already lowered a small boat with the avowed purpose of rowing to the *Philadelphia* whether a favorable or an unfavorable answer resulted.

A reply that sounded very much like "You come" came from the captured frigate.

Instantly Decatur leaped into the small boat with a rope, followed by a couple of his men.

In a few minutes they had pulled to the big shadowy hull of the *Philadelphia* and made fast their line to her forechains. He then twitched the rope sharply three times, a prearranged signal, and at once his crew on the ketch began to haul the latter toward the pirate craft.

At this juncture, the Tripolitans' suspicions seemed suddenly awakened. There came a harsh, alarmed cry of "Americanos!" and the strangers were ordered away in no uncertain terms in pidgin English.

But those buccaneers knew not with whom they were reckoning.

At the first sign of alarm on the part of the enemy, Decatur sprang up the main chains of the *Philadelphia*, shouting to his men to board her at once. He was rapidly followed by his officers and crew.

As the Americans swarmed over the rails and came upon the deck the pirates gathered in a panic-stricken, confused mass on the forecastle. Apparently. they thought themselves assailed by an opponent many times more numerous than themselves, whereas, in truth, the odds were all on their own side.

Decatur waited a moment until his men were all behind him. Then, placing himself at their head, he drew his sword and rushed upon the Tripolitans.

There was a sharp, but very brief conflict. The pirates, terrorized from the beginning, stood before the fierce onslaught only long enough to see scores of their number go down under the unerring pistol shots and cutlass thrusts of the Americans, and then those who could flee to the rails, jumped madly overboard.

Could he have done so it would have been the immediate effort of Decatur now to take the *Philadelphia* out of the harbor.

But this he knew to be impossible. He therefore gave orders to burn the ship, and his men, who had been thoroughly instructed in their parts, dispersed to all sections of the vessel with the combustibles that had been prepared. So well and quickly did they work, a few minutes later flames broke out in practically all parts of the frigate at once.

Satisfied that he could do no more, Decatur ordered his men to return to the *Intrepid*. Quietly, without confusion, they obeyed, although there was a pang in the throat of many a man of them because the dear old Yankee frigate must perish in this ignoble manner.

Burning of the *Philadelphia*, artist unknown.

It was a moment of unusual peril. The fire was breaking out all along the *Philadelphia*, and the flaming brands and sparks were already beginning to fly. Should any of these fall onto the deck of the little *Intrepid* nearby, she too would be afire in a moment, with great danger of her powder magazine catching and blowing her to fragments.

The Americans were quickly aboard their own craft, however. The next instant they had slashed the cables, manned the sweeps, and were bearing away from the larger ship.

It was a magnificent sight to see those great licking tongues of pallid red bursting forth from the hull and decking of the *Philadelphia*, as the ketch swiftly drew farther and farther away, afraid that in the bright and expansive reflection of the conflagration the enemy on shore could detect them and shoot them down before they could get out of range.

Soon, the tongues of fire were creeping up the tall masts of the frigate and clutching first one shivering shroud and then another in their consuming tentacles.

As the ship's guns became heated, they were discharged, as if by phantom gunners, shattering the silence of the night into quivering echoes.

And, as if in a final effort to wreak vengeance on the foreigners who had brought her to this, the grand old frigate now let go her shoreward batteries. The heavy shot went hurtling into the midst of pirate craft and shore works.

Finally her cables parted, and, an awesome pyre of wreckage, she drifted across the harbor to finally blow up.

Meantime the waters in the vicinity had been illuminated almost like day.

The escaping *Intrepid* stood out against the light background like a blot of ink on a white sheet of paper.

An angry roar went up from the buccaneer populace of Tripoli.

Shore guns and ship's guns began to cast their venomous hate toward the frail craft of the fleeing ones. But fortune favored the Americans; they kept on pulling for dear life. A friendly puff of wind filled their sail, and soon they were out of range and once more engulfed in darkness. A little later, they came up with the *Siren*, and bore away to rejoin Commodore Preble.

This successful venture was carried through solely by the cool courage of Stephen Decatur and the admirable discipline of his men.

The hazard was very great; everything had depended on the nerve with which the attack was made and the completeness of the surprise. Nothing miscarried; no success could have been more complete.

Lord Nelson—at that time in the Mediterranean, and the best judge of a naval exploit that ever lived—pronounced it "the most bold and daring act of the age."

In any event, it is one of those feats of arms which no American should ever forget, for although of no great importance, it fittingly illustrates the keen resourcefulness and high courage of American seamen.

CHAPTER VI

PERRY'S VICTORY ON LAKE ERIE

We Have Met the Enemy and They Are Ours, depiction of Perry's Famous Victory on Lake Erie in War 1812. Artist unknown.

They fought like tigers on that deck
Till ship was shatter'd and a wreck;
But as she sank in splendor to her grave
The brave old flag did still wave
From her last boat, speeding well,
'Midst burning shot and shrieking shell,
And fearless sailors pulled at oar
Till flag adorned a ship once more.

—*Atherton.*

IN THE EARLY 1800s, a serious disagreement broke out between the United States and Great Britain.

The issue centered around the British demanding the right to search any American vessels and seize any British naval deserters who had joined the American Merchant Marines.

America disagreed.

On June 18, 1812, just a few years after winning the right to become an independent country, the newly formed United States declared war on Britain.

American Commodore Oliver Hazard Perry was a 27-year-old native of Rhode Island. He assumed the command of Lake Erie and the other upper Great Lakes.

In the beginning, this position was a rather nominal [existing in name only] command because Lake Erie—at that time the most important waterway of all the lakes—was under exclusive control of the English foe.

But the Americans had captured a British brig called the *Caledonia* and purchased three schooners which were afterward named the *Somers*, the *Tigress*, and the *Ohio*. They also bought a sloop named the *Trippe*.

For a time, these ships were blockaded in the Niagara River. They were released only after the battle of Fort George which forced the retreat of the English forces.

By the hardest labor, Commodore Perry was able to tack these ships up against the current to Presque Isle (now called Erie), where two twenty-gun brigs were being constructed under his directions. Three schooners—the *Ariel*, the *Scorpion*, and the *Porcupine*—were also built.

The harbor of Erie was well-protected and spacious, a most excellent place for Perry's indefatigable efforts at building up a small fleet with which to chase the enemy of the waters under his direction.

But the harbor possessed one feature that, while an aid in keeping the British from getting in, now became a disadvantage to the Americans in keeping them from getting out.

The problem was an extensive sandbar that ran across the harbor's mouth. At times, the storms shifted this sand so much that vessels of deep draught [the depth of the vessel below the waterline] could not pass over it; at other times, the waters wore down the sandbar so the passage could be made quite readily.

But it was a dangerous, if not an impossible, obstacle.

British sailors leaving their ships and rowing toward a fort during the War of 1812.

While Perry was busy devising a method to circumvent the sandbar, Captain Robert Barclay, commander of the British forces on the lake, appeared off the harbor and proceeded to make matters even tighter for the Americans by instituting a blockade.

The enemy maintained the blockade until August 2nd, 1812, when their ships disappeared.

Anticipating an early return of the enemy, Perry made hasty preparations for escape.

On the 4th, at two o'clock, one American brig, the *Lawrence*, was towed to that point of the bar where the water was deepest. Her guns had been dismounted and placed on the beach, to make her as light as possible.

It was proposed to get the ship over by what is termed a "camel." This is how Commodore Perry himself, in a report, describes the procedure:

"Two large scows [flat-bottomed boats with broad square ends used for transporting bulk material], prepared for the purpose, were hauled alongside, and the work of lifting the brig proceeded as fast as possible. Pieces of massive timber had been run through the forward and after ports, and when the scows were sunk to the water's edge the ends of the timbers were blocked up, supported by these floating foundations. The plugs were now put in the scows, and the water was pumped out of them. By this process the brig was lifted quite two feet, though when she was got on the bar it was found that she still drew too much water. It became necessary, in consequence, to cover up everything, sink the scows anew, and block up the timbers afresh. This duty occupied the whole night."

Barely had the *Lawrence* passed the bar at eight o'clock the next morning than the enemy suddenly appeared.

The British were too late.

Captain Barclay exchanged a few shots with the American craft and then withdrew.

The *Niagara* was the next ship to cross the bar, then the others followed in succession. At first, there were not enough men to man the vessels, but a draft contingent arrived from Ontario, Canada, and many frontiersmen volunteered, while soldiers were also sent on board.

On the 18th this squadron sailed in quest of the enemy.

After cruising about some time, the *Ohio* was sent down the lake. The other ships went into Putin Bay.

British Captain Barclay found that he was very short of provisions. Up to this time he had not cared to risk an engagement with the Americans, but now he felt compelled to do so. Therefore, he prepared his vessels for action, and put out of Amherstburg on the 9th of September.

It was the next day that his squadron was discovered from the masthead of the *Lawrence*, just as the first streaks of dawn were appearing.

The British were then coming up from the northwest.

At once Perry's ships were all hustle and bustle, as it was realized that an engagement with the foe was almost sure to result. The light breeze that was blowing shifted to the northeast, thus giving the Americans the weather advantage.

In the meantime, Captain Barclay lay in a close column, heading toward the southwest. His ships, six in number, consisted of two men-of-war, two schooners, a brig, and a sloop. These were named respectively, and armed, thus: the *Detroit*, nineteen guns; the *Queen Charlotte*, seventeen guns; the *Lady Prevost*, thirteen guns; the *Hunter*, ten guns; the *Chippeway*, one gun; the *Little Belt*, three guns.

The American squadron, consisting of nine vessels, was made up of the brigs *Lawrence*, *Niagara*, and *Caledonia*; the schooners *Ariel, Scorpion, Somers, Porcupine,* and *Tigress;* and the sloop *Trippe.*

Their armament stood: *Lawrence,* twenty guns; *Niagara,* twenty guns; *Caledonia,* three guns; *Ariel,* four guns; *Scorpion,* two guns; *Somers,* two guns; *Porcupine, Tigress, Trippe,* each one gun.

In addition to having more ships than the enemy, the Americans had close to double armament strength, possessing fewer guns but guns of greater range.

Against this advantage, the British had an excess of perhaps fifty men in fighting trim. Therefore, it will be seen that if Commodore Perry had not come out victorious in the forthcoming battle, it would have been a national disgrace. It is true that his men fought with wonderful gallantry, but so did the men of Captain Barclay.

Battle of Lake Erie by artist Birch Thomas (1779–1851)

As the American squadron approached the foe, Perry's straggling line (the ships did not seem to keep well together) formed an angle of about fifteen degrees against the more compact line of the British.

At eleven twenty-five, the *Detroit* opened up the action by a shot from her long twenty-four, which fell short.

Ten minutes later she fired a second time. This was better directed and went crashing through the *Lawrence*.

Almost immediately, the Scorpion retaliated with her long thirty-two but missed.

The *Lawrence* shifted her port bow chaser, and now opened up herself with her long twelves, followed by her carronades [short smoothbore cast-iron cannons], but the shots all fell short.

The action now broke out generally, although the rearmost American vessels were almost beyond range of their own guns, and quite out of reach of the guns of their antagonists. The ships in the vanguard, however, were enough in numbers to match the enemy and keep him busy.

Meantime the *Lawrence*, crippled considerably by the shot that had struck her, bore down slowly upon the enemy.

It was twenty minutes before she succeeded in getting near enough to use her shorter-range guns—the carronades.

As she had approached she seemed to be the universal target for the foe, who directed their fire at her with gusto. Suffering a good deal, she was now able to use effectively all of her own armament. Her shots went into the ranks of the British line with telling vigor and accuracy, so much so, in fact, that the British speedily showed more or less distress.

Helping the *Lawrence* were the *Scorpion* and the *Ariel*. These three American ships, throwing a combined weight of one hundred and four pounds, were pitted against the *Chippeway* and *Detroit*, throwing one hundred and twenty-three pounds at a broadside.

At the same time, the *Caledonia*, the *Niagara*, and the *Somers* were engaging at long range, as were the *Hunter* and the *Queen Charlotte*, while the remaining American craft, the *Prevost* and the *Little Belt*, were also engaging at long range.

By twelve thirty, the *Lawrence* had worked into close quarters, and action was going on with the greatest fury between her and her antagonists within canister range.

Through ignorance, the raw and inexperienced American gunners committed the same error that the British so often fell into on the ocean during the war—they overloaded their carronades.

In consequence, the carronade of the *Scorpion* burst and kicked from its mountings in the middle of the action and went tumbling down the hatchway, while its shot scattered harmlessly against the hull of the *Detroit*.

One of the *Ariel*'s long twelves was also put out of commission by bursting.

During this time Captain Barclay's crew on the *Detroit* fought with excellent skill and bravery. Their aim was good, although the ship's equipment was so deficient that they actually had to discharge their pieces by firing pistols at the touchholes [the small hole in the cannon's rear that links to the ignition spark], being without slow matches.

Meanwhile the *Caledonia* came up into better range; but the *Niagara* signified no such intention, her commander, Captain Elliott, handling her wretchedly for the good of his compatriots, and keeping at a distance which prevented the effective use of his carronades on either his special antagonist, the *Queen Charlotte*, or the latter's sister ships.

Yet he had managed thus far to use his long guns well enough to cause the *Queen Charlotte* some damage, among which was the loss of her gallant commander, Captain Finnis, and Mr. Stokes, her first lieutenant.

The command had been taken over by Provincial Lieutenant Irvine who, perceiving that he could accomplish nothing in his present position, left off engagement with the diffident *Niagara,* passed the *Hunter,* and joined the attack on the *Lawrence* at close quarters.

Thus was the *Niagara,* the best-manned of the American vessels, kept out of action by her captain's misconduct!

At the end of the line, the fight went on at long range between the *Somers,* the *Tigress,* the *Porcupine,* and the *Trippe,* on the American side; and the *Lady Prevost* and the *Little Belt,* on the British side.

Considering her short-range equipment the *Lady Prevost* put up a remarkably stout and admirable defense against the long guns of her American adversaries. Before long she was necessarily greatly shot up; her commander, Lieutenant Buchan, had been dangerously wounded, and her acting first-lieutenant, Mr. Roulette, had also been seriously hurt. She began falling gradually to leeward.

Battle of Lake Erie illustration from an out-of-print book (artist unknown). The ships shown (left to right): *Little Belt* (British, 8 guns), *Lady Provost* (British, 18 guns), *Detroit* (British, 19 guns), *Chippeway* (British, 1 gun), *Queen Charlotte* (British, 17 guns), *Hunter* (British, 10 guns), *Lawrence* (American, 20 guns), *Scorpion* (American, 2 guns), *Caledonia* (American, 8 guns), *Ariel* (American, 4 guns), *Somers* (American, 4 guns), *Tripper* (American, 1 gun), *Porcupine* (American, 1 gun), *Niagara* (American, 20 guns), *Tigress* (American, 1 gun).

At the head of the line, the fighting was fierce and bloody to an extraordinary degree.

By now the *Lawrence, Scorpion, Ariel,* and *Caledonia* were opposed to the *Detroit*, the *Chippeway*, the *Queen Charlotte*, and the *Hunter*.

Lying at very close quarters, the ships of both sides did great destruction to their opponents, and their crews performed with a rare courage. Altogether it seemed quite an equal contest, the Americans being superior in weight of broadside metal, but inferior in men.

However, the *Lawrence* had received such damage in coming into the fight, that weighing her condition, the odds were somewhat against Commodore Perry.

As the men fought on, the heaviest fire on each side seemed to be concentrated upon the larger opposite vessels. The *Queen Charlotte* was now almost disabled, and the *Detroit* was battered and cut up in a fashion to greatly distress her. Indeed, her first lieutenant had fallen, mortally stricken, while Captain Barclay was so severely injured that he had to be carried from the deck, leaving the command to Lieutenant Inglis.

But if matters had gone ill with the *Detroit*, onboard the *Lawrence* affairs were even worse. Her decks showed evidence of the grimmest sort of carnage.

Out of one hundred and three men only nineteen remained unwounded. In her ward-room, used as a cockpit, and mostly above water, many of the wounded, in the hands of surgeons, were killed outright by random shots of the enemy which came hurtling through the thin wooden walls. Her first lieutenant, Mr. Yarnall, was three times wounded but kept the deck through all; the only other lieutenant on board, Mr. Brooks, was mortally wounded. Every brace was shot away; only one usable mast remained; her hull was a gaping sieve above the waterline; by degrees her guns on the engaged side had been all dismounted.

Perry, however, kept up the fight with praiseworthy zeal and unexampled bravery.

As his crew on the *Lawrence* fell one by one, he called down through the skylight for a surgeon's assistant; and this call was repeated and obeyed till no more assistants were able to answer. Then he shouted into the ward room: "Can any of the wounded pull a rope?" and three or four of them crawled up on deck to lend a feeble hand in placing the last guns.

It was the Commodore himself who, assisted by the purser [person on board who handles the money, or "purse"] and chaplain, aimed and fired the final shot.

With steady hand and the utmost coolness Perry sent his charge whistling toward the *Detroit,* and had the satisfaction of seeing a portion of her rail and foremast go flying into splinters.

Most commanders now would have struck their colors and permitted the enemy to take the crew prisoners.

But not so Perry.

Instead, the Commodore leaped into a rowboat with his brother and four seamen, taking his flag with him. Whatever happened, he had made up his mind that the British should never have a chance to gloat over *this* emblem—to handle its folds, and say when they got back to England, "This is one of the flags we took from the impertinent Americans."

The painting by William H. Powell (1865) hangs in the US Capitol. It shows Admiral Oliver Hazard Perry in a rowboat being transferred from the U.S. Brig *Lawrence* to the U.S. Brig *Niagara.* Note the flag being taken with Perry's men.

A quarter of a mile to windward, Perry's searching eye had already noted the *Niagara*—the *Niagara,* fresh, strong, still able to put up a good fight.

Now, standing proudly in the stern of the small boat, bareheaded, with the beloved flag flying defiantly from the bow, he ordered his men to pull hard for the *Niagara.*

It was two thirty when Perry arrived safely on board the *Niagara.* He transferred his flag, took command, and sent its former commander, Elliott, back in the rowboat to order up the three schooners. Within fifteen minutes the latter had come up, and Perry bore down to break Captain Barclay's line.

The British ships had fought themselves to a standstill. The *Lady Prevost* was crippled and sagged to the leeward, though ahead of the others. The *Detroit* and the *Queen Charlotte* were so disabled that they could not offer resistance against fresh combatants.

As the *Niagara* stood down, firing her port guns into the *Chippeway,* the *Little Belt,* the *Lady Prevost,* and her starboard guns into the *Detroit,* the *Queen Charlotte,* and the *Hunter,* she had everything her own way.

Raked terribly by her fire, the *Detroit* and the *Charlotte* tried to move off, but ran afoul of one another with their rigging shot to shreds.

Then the *Niagara* luffed smartly athwart their bows, within half pistol-shot, maintaining a terrific discharge of great guns and musketry.

On the other side, the British tangled ships were almost as disastrously swept by the *Caledonia* and the schooners, which were so close that some of their grapeshot [smaller-caliber round shots packed tightly, rather than a single solid projectile], passing over the foe, rattled against Perry's spars and braces.

Battle of Erie

CAUGHT AS HE was between two fires, there was nothing for Captain Barclay to do but strike his colors, which he did at three o'clock.

The *Chippeway* and the *Little Belt*, seeing the *Detroit* fall, did their best to escape, but were overtaken and brought to respectively by the *Trippe* and the *Scorpion*. The commander of the latter, Captain Stephen Champlin, fired the last shot, as he did the first, of the battle.

In its material results, as well as in its moral effects, Perry's victory on Lake Erie was most important.

It gave the Americans complete command of all the upper Great Lakes, and prevented continued fears of invasion from that quarter.

It also gave the British a greater respect for American naval ability, and surely did much to increase self-confidence in the nation.

CHAPTER VII

THE *CONSTITUTION* AND THE *GUERRIERE*

M. Corne p. A. Bowen, sc.
THE CONSTITUTION BEARING DOWN FOR THE GUERRIERE.

The queenly ship! brave hearts had striven,
And true ones died with her!
We saw her mighty cable riven Like floating gossamer;
We saw her proud flag struck that morn,
A star once o'er the seas—
Her anchor gone, her deck uptorn—
And sadder things than these!

—*Hemans.*

An Exciting Chase

THE FIRST CRUISE of the United States Navy in the War of 1812 was destined to be a disappointment to all concerned.

Commodore Rodgers set out with six men-of-war on June 21, 1812, three days after the declaration of hostilities. He finally returned to Boston harbor, after a seventy-day fruitless hunt for the British Jamaica squadron, with one letter-of-marque [government license in the Age of Sail authorizing a private person, known as a privateer or corsair, to attack and capture vessels of a nation at war with the license's issuer], seven merchantmen, and one recaptured American ship.

To add to the humiliation, an engagement had taken place with the British ship *Belvidera* of thirty-six guns, and she had very adroitly [look that word up—and how to pronounce it!] slipped away.

After her lucky escape, the *Belvidera* made her way to Halifax, Nova Scotia, Canada, the chief naval station of Great Britain on the American coast.

Her report was the first news of America's declaration of war to reach the British. News traveled slowly in those days.

Thoroughly alarmed, the English were quick to act, and in a few days, a squadron left Halifax in search of Commodore Rodgers.

The force thus hurriedly gathered was quite formidable. It consisted of: The *Africa,* sixty-four guns; the *Shannon,* thirty-eight; the *Guerriere,* thirty-eight; the *Belvidera,* thirty-six, and the *Æolus,* thirty-two.

It was the intention of the squadron to chastise the headstrong Americans for their attempt to dispute with Great Britain her proud mastery of the seas.

During the early part of July, this force made its appearance off New York and quickly effected captures enough to convince the American merchantmen that a season in port was preferable to the dangers of trading on the ocean in wartime.

The squadron also overhauled the American brig *Nautilus*, thereby gaining the distinction of taking the first war vessel of the war.

But what the British ached for was to meet enemy craft of more importance than merchantmen and fourteen-gun brigs. They must have been very pleased when, some days later, the United States frigate *Constitution* was in sight under circumstances that seemed to promise her easy prey for the five British warships.

It was on the 17th of July that the *Constitution*, after receiving a new crew at Annapolis, was standing northward under easy sail, bound for New York.

Along about noon, the lookout reported four sails on the horizon. One hour later a fifth sail was seen bearing up a few points to the eastward.

A careful scrutiny of the strangers convinced Captain Isaac Hull, the valiant commander, then thirty-seven years old, that they were men-of-war, although their nationality could not be determined. Nor was he able to make out their identity when night fell; while they were closer, they seemed not to be approaching him directly.

Uneasy and bent upon knowing just who they were before darkness prevented, Hull now set signals [either flares of short blasts from the ship's horn] that put the question plainly.

But although the distant ships must have seen and read these signals, they made no reply.

Then the night closed in earnest and they were blotted from view.

When day broke, to his chagrin Hull found himself fairly surrounded by British frigates!

Not only were there the original five British men-of-war that had left Halifax, but there was the captured *Nautilus* with her guns turned against the representative of her own nation, and a captured American schooner which had likewise been pressed into service.

Clearly the *Constitution* was outnumbered, and in a bad predicament.

Captain Hull saw that the only way he could save her was to take to flight.

Flight it was then!

The events of that three-day chase are told with great detail in the log book of the *Constitution*. To this record, many of the officers and crew have later added interesting and illuminating sidelights.

In referring to the early stages of the chase, British historians themselves freely admit that the *Constitution*, in taking advantage of the sudden lulls and gusts of the wind, showed far better seamanship and command than England's own vessels.

Later on, the frustrated foe called in all their small boats, and when the American frigate had vanished from view, went about for some days picking up her ships that had drifted widely apart.

To the delight of the sailors aboard the *Constitution*, the morning of the second day brought a light breeze over the ruffled waters which promised to hold. It really did keep fairly steady for a few hours, during which time she gained on her pursuers sufficiently to put her in the lead close to five miles.

Then the breeze died out and the calm again held them in its grip, calling for the sweeps and small boats and kedge [light anchor].

Painting depicting the boats of USS *Constitution* towing her in a calm, while she was being pursued by a squadron of British warships, 18 July 1812. Unknown artist.

On the gun-deck, about the carriages of the great cannons now useless, lay the crew that was not assigned to duty in the boats or at the capstan [a vertical spool-shaped cylinder used for hoisting weights].

Wearied with the constant strain, they fell asleep as soon as relieved from active duty, although they knew that from that slumber they might be awakened in the midst of yelling comrades, the clash of steel, and the roar of guns.

Ever alert for any sign of a coming breeze, Captain Hull saw on the water far to windward a rippling appearance that indicated wind. He determined to utilize this change in such a way that the British themselves would gain nothing by it. Clouds, too, began to appear to the windward side, promising a squall of uncertain proportions. In a short time, sheets of gray, which the experienced sea-dog knows to be rain, could be seen driving toward them.

The *Constitution* was made ready as if she expected a severe gale.

The enemy could see the nimble American sailors taking in sail and furling all the lighter canvas. They could also see the clouds and rain. There was no doubt about it; a bad squall was coming.

Immediately all was activity in the tops of the British frigates. Knowing the peril of being caught in a storm too close together, the British warships steered well apart; the course was forgotten in the effort of every shipmaster to meet the brunt of ill weather as best he could.

Before these preparations had been completed the rain had reached the *Constitution*, and she was enveloped in a heavy shower which so hid her from view of the enemy that her outline was all but lost to the British.

For this reason, the foe was not aware that, in the midst of the deluge of rain, the American sailors were tearing around and shaking out the sails to almost full spread [more sail area equals more power for the ships that are fueled by wind].

Then, with the uprising wind filling her canvas into beautiful arcs, the *Constitution* sprang forward like a hound after a rabbit, headed for Boston harbor, the seas dashing high against her sharp bow and at times wetting the fore deck.

After traveling thus through the storm, which lasted upwards of an hour, the thunder clouds passed by, and with them went the rain to windward. But, to the joy of the American sailors, the wind still held good, and their ship continued to drive well.

As the gray curtain of rain swept farther and farther away, the eyes of the sailors followed it with increasing intentness and exultation. As more and more blue water was disclosed in the direction of the British, without showing a ship, many a man said to his neighbor with a chuckle, "Where are they?"

At last "they" appeared.

So far were the British left behind that they were now practically out of the chase. Two of them were actually hull-down, while one was a mere speck against the horizon-line.

Blanketed by the storm, they had not seen their shrewd fugitives getting away, had made no effort to throw their furled sail until the squall had passed, and now they must have been immensely crestfallen at the trick played upon them so cleverly.

But though far behind, the foe must be given credit for not giving up the pursuit. They held valiantly after the American frigate, hoping against hope to yet overhaul her.

It was a fruitless effort.

Instead of gaining on the *Constitution*—which was really a swifter craft—the British ships fell farther and farther behind. Finally they had all disappeared from view, and the American frigate continued on her way to Boston without mishap.

Guerriere and Constitution.

An Unexpected Meeting

"YOU SAY HER name is the *Guerriere*?"

"Yes; I have just recently been commissioned to her. She carries thirty-eight guns and is as fast as a hound. A fine craft, Hull!"

"Mayhap she is, Dacres; but you may just watch out for that ship of yours if I ever catch her in the *Constitution*!"

Both speakers were dressed in the uniform of naval officers.

One—the taller—wore the British insignia of a captain; the other, the markings of an American captain. The Englishman now laughed good-naturedly.

"If war should come, Hull, I hope two such friends as we may be spared the irony of meeting one another in conflict; but if so, I am willing to bet a hundred pounds sterling that your *Constitution* will strike her colors to my *Guerriere*! What say?"

"No," said Hull; "I'll bet no money on it, and could not use your English sterling [money of the time] anyhow. However, I will stake you a hat on the outcome in favor of the *Constitution*."

"Done!" responded Dacres promptly; and the bet was made.

This little banter ended a pleasant discussion at a social gathering in Philadelphia while the vessels of the two officers of different nations lay in the Delaware. Although the cloud of war hung threateningly over the edge of the horizon, old friends Captain Hull, an American, and Captain Dacres, an Englishman, would not allow bitterness to creep into their hearts against each other.

USS *Constitution* fires into the *Guerriere*. Painting by Anton Otto Fischer, date unknown.

How little did either man realize that they were very soon to be pitted against each other on two different occasions—one in a stern chase of many miles, when neither dreamed of the identity of the other—the other in ferocious combat, each man squarely set, each forgetting self and friend and thinking only of duty to country!

Such bitter incidents are common to the demands of war.

We have already seen, in the preceding chapter, the result of the first meeting of these two friends, now rivals.

We shall now see how they finally met by accident and forever settled the question of which of their proud ships was the better fighter.

After her escape from the British fleet, the *Constitution* remained at Boston only a few days, and then set out on a cruise to the eastward along the New England coast.

Bad luck seemed to follow her. Not a prize did she take until she reached a point off Cape Sable. Here she captured two small English merchantmen and forced an English sloop-of-war to relinquish an American brig.

Shortly afterward a Salem, Massachusetts, privateer informed Captain Hull that he had seen a British frigate cruising in the neighborhood.

This was good news.

If Hull could only come up with the foe warship, the monotony of the cruise might be broken. He straightway set out to scour the sea in quest of the frigate. Never once did he dream that it was the *Guerriere*—the ship of his pre-war friend, Captain Dacres!

It was not long before the search was ended.

One morning there came a long drawn hail of "Sail ho-o-o!" from the lookout aloft.

The course of the *Constitution* was at once shaped toward the stranger, who was not yet visible from the deck.

In half an hour she was made out to be a frigate. As she was seen to alter presently her own course and bear toward the American vessel, it was assumed that she was either a friend or a very audacious foe.

Without waiting longer, determined to be on the safe side, Captain Hull began operations for action. The top-gallant sails were furled, and the lighter spars lowered to the deck. The decks were cleared, and the guns uncovered.

Through their glasses, the officers could see the stranger making similar preparations; still, it was uncertain that she was an enemy.

By five o'clock in the afternoon, the two ships were rapidly nearing, but were still too far off to accurately identify.

The drums on the American frigate beat to quarters. Then followed the rush of barefooted men along the deck, as they ran hastily and in perfect order to their various stations. As the roll of the drums died away, the shrill voices of the midshipmen rose, calling off the quarter bills, answered by the gruff responses of the men at their posts.

The *Guerriere* was not one bit behind in her preparations for a possible conflict.

Although Captain Dacres had been unable to establish the identity of the other vessel, he was too good a commander, like Hull, to take it for granted that she was a friend.

He called to his side Captain William Orne, an American sailor whom he had taken prisoner with Orne's vessel some days before.

Handing the American his glass, Captain Dacres said:

"Captain Orne, what do you think of that vessel? Is she friend or foe to me?"

Taking the spyglass, Orne inspected the distant frigate carefully. He saw by her peculiar sails and general appearance that she was without doubt an American vessel. Being too honest to deceive the British captain, as he might have done to the advantage of his friends had he chosen, he answered candidly:

"I think, sir, from her behavior, that she is an American frigate, but I do not recognize her."

"She has a familiar look to me somehow," said Dacres, puzzled. "I can hardly credit her with being an American, however, for she comes down too boldly for such."

"It is by her very boldness that I make certain she is an American," retorted the prisoner with great warmth.

The British officer laughed good-humoredly. "So be it, sir," he said. "The better she behaves, the more credit we shall gain by taking her."

As the two ships came down toward each other, the *Guerriere* backed her main topsail and waited for her opponent to draw near enough to commence action—if action there was to be. As Dacres stood thus, he set the English flag at each masthead, and beat to quarters.

When the frigate came to within two or three miles, she took in all her light sails, reefed [shortened] her topsails, and made final preparations for combat. Then she filled her sails and ran down to meet the *Guerriere*.

On board the latter, Captain Dacres turned again to his American prisoner.

"Captain Orne," he said politely, "as I suppose you do not wish to fight against your own countrymen, you are at liberty to retire below the waterline."

The American officer bowed, filled with admiration for the gallant conduct of the British captain, and started for the cockpit. With him went half a dozen American seamen whom Dacres had drafted into his crew, and whom he now chivalrously relieved from duty.

Such fine conduct is not often met in the home of an enemy.

Now Captain Dacres saw the Stars-and-Stripes go up to the masthead of the frigate and realized that his preparations had been well-timed.

Instantly he let go with his broadside. Observing that his shot had fallen short, he wore the *Guerriere* around and tried her port broadside. Most of this went through the American's rigging, though two shots took effect in her hull.

In response to this, the *Constitution* yawed a little.

Bang! bang! went two of her bow guns.

Once more the *Guerriere* fired her broadsides.

In this manner the battle continued for about an hour at long range, the American ship saving her ammunition and only responding to the heavy broadsides of her antagonist with occasional shot.

As this ineffectual firing continued, the two vessels had been slowly drawing nearer and nearer, and the gunners on the *Constitution* became so restless under their inaction that they could hardly keep still.

Captain Hull was pacing the quarter-deck with short, quick steps, trying to keep cool, but inwardly much excited.

While thus engaged, Lieutenant Morris, the second in command, came up and asked his permission to let him respond with a broadside, declaring the gunners were becoming almost unruly in their impatience to do more firing.

But Hull shook his head. Some minutes later the request was repeated.

This time there was no response at all; Captain Hull seemed too engrossed in pacing to and fro and watching the enemy to even hear the words or see his junior officer, who, nonplussed, retired once more.

When, a little later, the ships were within half-pistol-shot of one another, the smothered excitement in Hull's breast suddenly broke out.

"Now, boys, pour it into them!" he shouted at the top of his lungs, gesticulating with such violence that the tight breeches of his naval uniform split far down the side.

Lieutenant Morris repeated his superior's orders. "Hull her, lads! Hull her!" he cried.

The crew joyously yelled in chorus: "Hull her! Hull her! Hull her!"

The guns had already been carefully aimed. Now, with slow-match, they were touched off. The effect of the first broadside was terrific.

Captain Orne, deep in the cockpit of the *Guerriere*, said afterward that he "heard a tremendous explosion from the opposing frigate. The effect of her shot seemed to make the *Guerriere* reel and tremble as though she had received the shock of an earthquake. Immediately after this I heard a tremendous shock on deck, and was told that the mizzenmast was shot away. In a few moments the cockpit where I was was filled with wounded men. I did not like to see them suffer, but I was overjoyed that the American was getting in such good work."

THE CONSTITUTION AND GUERRIERE.

But in the seclusion of the cockpit, the American prisoner could gain little idea of the work of destruction going on above.

He could not see the gunners on the two ships, their bare breasts covered with stains of powder and rivulets of blood and sweat, pulling fiercely at the gun tackle and wielding rammers with frantic energy.

He could not see the British ship's mizzenmast go crashing and hurtling from its ragged stump into the sea, carrying with it some of the topmen and throwing others far out into the churning waters, choked with freshly made debris.

He could not see the great, ragged breach torn in the *Guerriere*'s quarter deck by the falling mast, nor observe the frantic energy with which the British sailors went to work to clear away the wreckage on deck.

While this was going on, the *Constitution* drew slowly ahead, pouring in several other destructive broadsides. Then she lulled until she lay right athwart the enemy's bow.

This brought the bowsprit of the *Guerriere* across the quarter-deck of the American, where it was soon fouled in the mizzen-rigging of the latter.

High and clear on the evening air now sounded the notes of two bugles. One came from the *Constitution*, the other from the *Guerriere*. Each was calling up its crew of boarders.

Like modern rival football gladiators, ready for the game, they responded to the signal. But instead of a trophy of goals to lead them to victory, in this case it was to be a trophy of dead men. Now they came rushing—boarding-caps on head, cutlass in one hand, cocked pistol in the other—to the near side of their respective vessels!

But a heavy sea is rolling and tossing the two frigates; to board either seems almost impossible.

As Captain Dacres observes this he recalls his men back to the guns.

Even though each party is forced to stick to its own ship, so close are they that all weapons except the cutlasses can be used.

From the tops comes a steady rain of leaden missiles down upon the heads of the adversary, one of which slightly wounds Captain Dacres.

The protruding muzzles of the big guns often touch the near side of the rival ship; when they are drawn in for reloading, after their deep-toned thunder, the sailors quickly thrust their muskets through the ports.

The rattle of small arms is almost incessant, sounding like millions of stones being cast against a suspended piece of sheet-iron; muskets and pistols blend together in their reports. And interspersed throughout it all are the ever-present human tones—sometimes in command, sometimes in anger, sometimes in reproach, sometimes in pain, sometimes in exultation, sometimes in sacrificial glorification, sometimes in weakened gasps—seldom in fear.

In the midst of the turmoil of heroic tragedy there suddenly shrills high above all other noises, a new human cry—a cry of portentous dread: "Fire! Fire! Fire!"

And, for a moment ceasing their fighting, the startled sailors aboard the *Constitution* see one of their own men pointing excitedly to the cabin, from the windows of which billows of smoke are rolling.

The fire has been set by the flash of the enemy's gun, so close are the ships.

By the hardest kind of exertions, men extinguish the fire. Then the fighting is resumed with renewed doggedness.

A few moments later, a grim American gunner manages to disable the enemy gun that has done the mischief. A cheer goes up from those near him; those in other quarters are too busy even to note the swelling cry.

Lieutenant Morris, with his own hands, is trying to lash the two heaving ships together. Giving up the attempt, he leaps to the taffrail and calls upon his men to follow him aboard the foe.

Before the last syllable dies on his lips, Lieutenant Bush of the marines, and Mr. Alwyn, are by his side.

Crash-h-h! goes a volley of British musketry, aimed directly at the intrepid three. All pitch backward—Bush dead, Morris and Alwyn badly wounded.

Now comes a ripping, crunching, rending of wood as the sea tosses the fouled ships and their entangled rigging tears asunder.

They slowly drift apart, free once more. The heavy smoke of the guns rolls in between, shielding both for a few minutes.

As it rises again, the *Constitution*'s big guns roar out once more; a great cheer arises from her decks as the *Guerriere*'s foremast is vitally fractured, topples, hangs uncertainly for an instant, then with a groan almost human, crashes down, carrying with it the mainmast and a score of unfortunate topmen.

The shattered British ship now lies a shapeless mass, tossing unguided upon the waves. She has lost all her backbone. Even her ensign flutters from a slivered stump.

Drawing away and firing continually and relentlessly with her stern guns, the *Constitution* presently maneuvers herself into a good raking position. As she is on the point of letting go her broadside, the colors on the enemy ship are hauled down.

What a cheer goes up from the Americans!—what a hoarse, tired, happy, boisterous cheer!

Captain Hull's rugged face bears a peculiar look. In his features, you read a strange mixture—triumph and pain; triumph for his beloved country, pain for his dear adversary. He cannot face him right now. Therefore, he calls Lieutenant Read and sends him aboard the *Guerriere*.

"Captain Hull presents his compliments, sir, and wishes to know if you have really struck your flag," states Read.

The Surrender of Captain Dacres, artist unknown.

Captain Dacres's face is a study. Then he responds dryly, with a significant glance at the shattered masts and bloody deck of his ship: "Well, I don't know. Our mizzenmast is gone, our mainmast is gone, our men are gone—and I think, on the whole, you may say that we have struck our colors."

After looking about the ship, which he found in a fearful condition, with scores of killed and wounded, untackled guns surging from side to side, and some petty officers and seamen even intoxicated, Lieutenant Read returned to the British captain.

"Would you like the assistance of a surgeon, sir, or a surgeon's mate, in caring for your wounded?"

Dacres looked surprised, and responded:

"I should suppose you had on board your own ship business enough for all your medical officers."

"Sir," answered Read, "we have only seven wounded, and they have been dressed long ago!"

Captain Dacres was astounded, as well he might be; for on the decks of his own ship lay twenty-three dead or mortally wounded men, while in the cockpit the surgeons were doing their best to alleviate the sufferings of fifty-six others.

The Americans now set to work to remove the prisoners from the *Guerriere*, which was evidently in a sinking condition. Needless to say, they were pleased to find that in the capture they had given freedom to some of their own countrymen—Captain Orne in particular.

In the first boat load from the British ship came its sad commander, Captain Dacres, who was politely shown into Captain Hull's cabin.

Unclasping his sword, the conquered British officer extended it silently and formally to his old-time friend.

But Captain Hull, with trembling lip, shoved it gently back.

"No, no, Captain," said he, "I'll not take a sword from one who knows how to handle it so courageously"—into his eyes came a roguish look—"but, if you don't mind, I will now trouble you for that hat!"

For a moment a shade of perplexity passed over the brow of the humbled British captain; then he recollected the wager of a year or two back, there in Philadelphia, and sheathing his formality with his sword, he shook hands laughingly but pathetically with his former comrade.

CHAPTER VIII

THE SHIP THAT STRANGELY DISAPPEARED

Marines aboard USS *Wasp* engage HMS *Reindeer*, June 1814. Painting by John Clymer (1907–1989).

What cruel fate is this,
That one so gallant, a master of the seas,
Fearing naught, sailing where she please,
Should vanish in the dark of night,
And never leave a sign in sight
No bit of wreckage, no floating mast
To show which way she might have pass'd?

—*Bellwright*.

FOR AS LONG as one hundred and fifty years before the War of 1812, England had been a virtual "mistress of the seas."

The warships of no nation were able to compete successfully against those of Great Britain, although, at that time, practically every European power had matched its naval craft against her.

And when, in 1812, the unknown navy of the new nation growing up across the Atlantic did what no old navy had ever been able to do, not only the English but the people of Continental Europe opened their eyes wide in wonder, and the former were greatly humiliated and aroused.

So, the Americans first proved to the world that the English could be beaten at their favorite game.

It was the Americans who did what the huge fleets of France, Spain, and Holland had failed to do. And—greatest credit of all—they did it without large fleets and squadrons, with a mere handful of frigates and sloops, hammered together at short notice, and manned and equipped with crude material.

But untrained as they were, those seamen were full of American spirit and initiative—so full of it that their deeds with solitary ships have been honored with as much attention by modern writers of naval warfare in the Old World as bestowed upon the actions of whole fleets in other wars.

Among the famous ships of the Americans in this war were two which carried the name of *Wasp*.

The first one built was an eighteen-gun sloop, which, at the very outset of hostilities, captured a British brig of twenty guns, but unfortunately, she was herself taken a little later on by an enemy.

In memory of her, the Americans gave the same name to a new sloop which they had been building.

These sloops of war were very stoutly made, and their swiftness compared favorably with any European ships of their class, for the American shipwrights were already as famous as the American gunners and seamen.

The new *Wasp*, like most of her sister ships of sloop build, carried twenty-two guns. She was ship-rigged, and had a crew of one hundred and seventy men. Twenty of her guns were carronades, shooting thirty-two-pound projectiles, and intended for short range. Her remaining two guns were bow-chasers, termed "Long Toms," which had far stronger carrying power.

The USS *Wasp*

It was in 1814, during the last year of the war, that the *Wasp* sailed from the United States to prey on the navy and commerce of Great Britain.

For commander, she had a gallant South Carolinian named Captain Johnson Blakeley. Her crew were nearly all Americans, being an exceptionally fine set of men, quick to act, and afraid of nothing above or below the sun except the wrath of God.

Instead of staying near the American coastline, or sailing the high seas for her prizes, the *Wasp* at once boldly turned her nose toward the English Channel, bent upon carrying the war to the very doors of the enemy, as John Paul Jones had done thirty-seven years before.

By this time, the British fleets had obtained such complete supremacy over European waters that the ships of the French, who were friendly with the United States, were almost completely bottled up, and could not get out to render the aid to America they would so much have liked to.

Night and day, in calm, squall, and tempest, these great blockading squadrons of England keep watch upon the rival warships of the French emperor, as a cat crouches and awaits the first movement of the little mouse for a quick pounce.

Other British ships-of-war patrolled the seas unchecked, ready to pick up the first rash French vessel or American ship they could see.

In spite of all this vigilance of the enemy, he found he could not close up all the gates into the seas, nor frighten those he watched into tamely staying off the waters. A few French privateers slipped out now and then. The bolder and more formidable American privateersmen drove hither and thither across the ocean in their swift craft, laughing at the enemy's frantic efforts to catch them, and harassing the English commerce without mercy.

Of these American privateers the *Wasp* was one of the most audacious.

She proceeded at once, upon crossing the Atlantic, to cruise boldly in the English Channel and off the coasts of England,

France, and Spain. Here, in the very teeth of British naval power, menaced by enemy squadrons guarding British merchantmen, imperiled by diverse foe ships-of-the-line, out looking for stray privateers like herself, the *Wasp* kept on.

Many was the time that her escape from capture was narrow; but by the splendid seamanship of her crew and the vigilance and skill of her commander, she kept on threading the dangerous home waters of her adversary, as saucy and daring as you please.

Before she had been long on the ground, one June morning found her giving chase to a couple of English merchantmen. In the midst of the exciting pursuit, her lookout from the masthead reported a strange vessel of suspicious character bearing down fast from leeward. A little later, the stranger was close enough for Captain Blakeley to see that she was a British brig-of-the-line.

She was, in fact, the *Reindeer*, of eighteen guns and one hundred and twenty men, commanded by Captain Manners of the Royal Navy.

USS *Wasp* (left) and HMS *Reindeer*. Engraving by Abel Bowen, 1838.

At once the American captain decided to drop his pursuit of the merchantmen and engage the warship.

The sky was a beautiful azure blue; the air mild and still. The very lightest wind stirred across the wide expanse of sea.

At one o'clock, the *Wasp*'s drum beat to quarters [the announcement aboard a naval warship that signals all hands must go to battle stations as quickly as possible], and at once, every man ran quickly to his post, while the sails were furled into fighting trim.

The *Reindeer*'s drum also beat to order, and she too made preparations for battle which could be plainly seen from the deck of the American vessel. On her forecastle, she had rigged a light carronade, and as she now came down from astern of the *Wasp,* she fired this gun five times point-blank at the American sloop, some of the shots taking effect in the latter's hull but doing little damage.

In answer, the *Wasp* now lulled slowly round, firing her heavy carronade as she bore.

In a few moments, the rival ships had closed; yard-arm to yard-arm they lay, almost in a mortal embrace. [Yard-arm is the outermost end of a horizontal bar that supports and controls the sail.]

Terrific was the thunder of the heavy guns of both. No sooner did a gun spit out its venom of hot lead than its grimy gunner pulled in its smoking muzzle, its equally grimy swabber swabbed it, and it was loaded and thrust through the port again for another merciless discharge of hate.

Like demons, these men worked at the breeches, while on the decks and the tops other men fired with pistols and muskets at every human target presenting itself.

As the vessels ground together, the American sailors sprang to the bulwarks [protective boundary on a ship's side to keep crews from falling into the sea] to make the union sure by throwing grapnels [a many-pronged anchor that could hook a ship to another] over the *Reindeer*.

They were protected by their brother sailors below and aloft, who discharged a veritable shower of shot into the foe, many of whom made desperate efforts to rush forward and frustrate by savage thrust of pike and bark of pistol and slash of cutlass.

In this struggle Captain Manners himself was wounded.

But a wound could not deter the brave British commander.

At this juncture, he gave his own men the order to gather for boarding.

Immediately, they came running up, many naked above the waist, their skin already streaked with sickening red and blacked with powder smudges, cutlass gripped in one rough fist, pistol in the other.

However, the Americans were quick to see their new danger.

Behind the bulwarks of the *Wasp* crouched the American pikemen, with set countenances and restless feet and hands; behind them, drawn up on the deck, cutlassed and pistoled, stood the sturdy, tanned marines of the new country; at the right and left, on stations of elevated vantage where they could note every movement of the foe, were the cool blue-coated officers, swords in hand, ready to meet any emergency themselves.

Now came Captain Manners's signal to his men: "Boarders away!"

Like a snapping bowstring, the taut nerves of the men on both sides flew into action.

The British sea-dogs began to tumble over the rail of the American ship—only to perish by shot or steel.

Still they persisted, brave men that they were. Desperately they stabbed with pike and cut arcs with their cutlasses. Fully as desperately did they pull the trigger of a pistol and attempt to force their way well upon the American deck.

Captain of the *Reindeer* commands his crew boarding the *Wasp*. Artist unknown.

Seeing his men stagger and hesitate, on the very verge of rout, the dauntless Manners shouted encouragement and sprang to their head.

Too late!

Even while his arm was extended forward, indicating the goal, the bullet of an American sharpshooter in the foretop came unerringly down and crushed through his skull.

As Manners fell, his men became panic-stricken.

They were quickly forced back to their own deck and then came the order of Captain Blakeley himself to board.

Over the bulwarks of the enemy ship they scrambled, shouting like men gone crazy. Absolutely irresistible was their advance. With wild cheers, they swept the wreck of the British seamen before them, and in almost the time it takes to tell it, the *Reindeer* was in their possession.

Every officer on board the enemy craft had been killed, while at least two-thirds of her crew had been killed or wounded. That they had fought valiantly was shown by the fact that twenty-six Americans had suffered death or injuries.

Having no desire to return to home shores right away, and not wishing to be handicapped with a prize consort, the crew of the *Wasp* removed the British to their own vessel and then applied the torch to the *Reindeer*.

After running into a friendly French port to refit, the *Wasp* once more put to sea in quest of new laurels. For some time she met no antagonist worthy of her timber; moreover, she had to exercise the utmost vigilance to escape capture.

Late one September afternoon she found herself within striking distance of an isolated British brig which belonged to a squadron whose other ships had become somewhat widely separated in a recent storm.

Although the brig's sister ships were too close to promise a successful attack on her in broad daylight, Captain Blakeley thought he stood some chance under cover of darkness.

Keeping the *Wasp* far behind, where her identity was unlikely even to be suspected, he trailed after the *Avon*, which carried eighteen guns and was in some respects a more powerful craft than the American vessel.

During the night he came up with the pursued after following his lights for some time, and bearing very close fired a heavy broadside into the *Avon*, whose astonished captain had taken him for one of the other British ships.

As soon as she could do so the *Avon* turned her own guns upon her attacker, while sailors set lamp signals high on her foremast calling to her compatriots for aid. Plunging and wallowing in the sea, which was running quite high, both sides found the aiming difficult. But the firing was fast and furious, nevertheless.

HMS *Avon* in full sail, by artist William Frederick Mitchell (1845–1914)

The British marksmanship was very bad; few hits were scored. On the other hand, the Americans displayed their customary superiority at this game, and before long with such effect that the hull of the British brig was in deplorable condition.

Noting that she was sinking, her commander struck his flag and cried for quarter. Fifty of his men had been laid low, while only three Americans had sustained wounds.

Before the *Wasp* could take possession of her prize, two British ships which had seen the signals of their distressed friend, or heard the firing, unexpectedly put in an appearance.

Extinguishing his own lights, Captain Blakeley adroitly sneaked away in the darkness, with one enemy shot chasing after him.

Had only one foe ship come to the rescue of the *Avon* he would promptly have attacked it.

With the enemy all around him, his feat of running in and destroying one of their ships was remarkably bold and wonderful in its accomplishment; he was too sagacious [wise] a seaman to spoil his good work now by remaining to engage insurmountable [look that one up!] odds.

After this, the *Wasp* took other prizes.

Once she came across a convoy of British ships carrying arms and munitions of war to Wellington's army [Duke of Wellington, a famous British military leader who defeated Napoleon at the Battle of Waterloo], under the protection of a big two-decker warship.

Hanging on the outskirts of this convoy, like a hawk over a barnyard, the swift sloop watched her chance and finally swooped down upon one of the transports and destroyed it, before the mothering ship could bring her guns to bear.

This was the last known feat ever performed by the gallant little *Wasp*.

Her country never again heard from her; Europe never heard from her.

Where had she gone?

The seas were scoured for her by friend and foe alike, but the waters seemed to have opened up and swallowed the daring sloop and her daring crew, leaving no human survivor, not even an identifying splinter of wood, scrap of metal, or fragment of uniform.

And nothing more is known about her to this day....

Here is the exciting end of Volume I of The Boys Book of Sea Fights.

This volume also makes the end of warships that were propelled by sails in battle.

Now the action REALLY ramps up!

Volume II will introduce you to the Ironclads—steam-powered ships that were equipped with much heavier guns and huge turrets. These modern ships could also fire explosive shells that could hit an enemy that was MILES away.

Volume II will also introduce you to the dreadnoughts (which means "fear not"). These were the first all-steel battleships, and they dramatically changed warfare forever.

So brace yourself and stand by for action in Volume II of The Boys Book of Sea Fights!

Made in the USA
Las Vegas, NV
02 December 2025

35228871R00105